T3-BRK-855

Waste Age/ Recycling Times'

RECYCLING HANDBOOK

Edited by
John T. Aquino
Editor and Publishing Director
Waste Age Publications
Environmental Industry Associations

Library of Congress Cataloging-in-Publication Data

Waste age/Recycling times' recycling handbook / compiled by John T. Aquino
p. cm.
Includes bibliographical references and index.
ISBN 1-56670-068-X
1. Recycling (wastes, etc.)—United States.
I. Aquino, John. II. Waste age. III. Recycling times.
TD794.5.W347 1995
363.72′82′0973—dc20 95-14479
CIP

Direct all inquiries to CRC Press, Inc., 2000 Corporate Blvd., N.W., Boca Raton, Florida 33431.

Lewis Publishers is an imprint of CRC Press

International Standard Book Number 1-56670-068-X
Library of Congress Card Number 95-14479
Printed in the United States of America 1 2 3 4 5 6 7 8 9 0
Printed on acid-free paper

PREFACE

In the Middle Ages, alchemists tried to turn base metals into gold. In recent times, newspaper articles have turned into a cliché the idea that recycling—processing material for reuse—is a modern form of alchemy, turning waste into gold.

Recycling is valued as the equivalent of gold in some cultures. In Mexico, for example, as in other Latin and South American countries, the *pepenadores* have lived for years near landfills and, by birthright, picked certain materials from the trash to sell for some form of reuse.

Nor is recycling new. The Romans who pulled the stones from the Coliseum to pave highways were pioneers in C&D (construction and demolition) recycling. The Germanic hero Siegfried, as portrayed in Richard Wagner's operatic *Ring* cycle, rather than patching his great sword, melted it down and reforged it. He was practicing metal recycling. The steel industry in America has almost always recycled its old steel.

What is new in the late twentieth century is the fervor with which recycling has been adopted. Many of the United States set a certain level of recycling. By focusing on package recycling, Germany has reportedly brought the recycling level for packaging to 50%, due to mandated packaging recycling. Similar mandates have been enacted in other European countries and even proposed in the United States. A hauling company I toured in 1992 told me of having acquired the municipal contract to pick up recyclables and being told the participation rate would be no more than 28%. The first day, over 80% of the community participated, and the haulers, who didn't have enough vehicles, made numerous trips and spent the evening knee deep in plastic bottles.

The fervor is due to a concern about the environment, about conserving our resources, and about the lack of landfill space in at least some of the United States and many other countries. Recycling is easy to understand—something new is made out of something old. It is not thrown away, it is reused.

And yet, the late 1980s, as the recycling flame burned bright, produced stories of gluts of some materials that had been picked up for recycling. Warehouses were full of old newspaper (ONP) that, rather than being sold for a profit, cost money to have removed. Market prices dropped. The hauling away of trash, which once was a direct trip to a landfill, now included the cost of the collection and marketing of recyclables. Hauling companies used to ask me to publish articles telling "the truth about recycling," which basically meant that it was unprofitable. While public interest in recycling was high, some companies wished that it would go away.

As this is being written, the markets for many recyclables are high. Technology has improved, decreasing some costs. Lessons that have been learned are that (1) modern recycling is still in the development stage, and the learning process will continue; (2) modern recycling is not going to go away; and (3) modern recycling is not the modern alchemy. It is not magic. It is a technological process that will be refined and improved as years go by.

Waste Age magazine is credited by many as having given the modern concept of recycling special exposure in the United States under the editorship of Joe Salimando in the mid- to late-1980s. In 1989, *Waste Age* introduced a sister publication, *Recycling Times.* It seems appropriate, then, to base this *Recycling Handbook* on articles that appeared in *Waste Age* in the early 1990s, sometimes reworked and expanded from *Recycling Times.*

In doing an anthology of articles, I am reminded of the story of Irish playwright and author George Bernard Shaw, who, when asked to republish his music criticism of previous years in book form, said that nobody was interested in reading about dead sopranos. However, the articles here have not been selected for their historical value, but rather as still pertinent explications of recycling issues, written for a professional audience. When appropriate, they have been updated.

The ultimate purpose of the book is to provide a detailed background of markets, regulations, basic workings, important issues, and representative programs related to recycling. The original audience for the articles was the audience of *Waste Age*—solid waste professionals—in both private companies and government—as well as engineers and consultants. In this revised and updated anthology format, the material in the chapters is intended for solid waste professionals to use as a resource and/or to develop a holistic, broader understanding of recycling, as well as for those who work with them or have to address the issues that the professionals address—attorneys, accountants, businesses interested in recycling or required to recycle, and the general public who want to understand the issues from the professional perspective.

—John T. Aquino

ACKNOWLEDGMENTS

Thanks must be given to Joe Salimando, *Waste Age* editor from 1985 to 1990, for his pioneer work in bringing the concept of recycling to more and more people; to Kathleen Meade, editor of *Recycling Times* from 1990 to 1992; to Cheryl McAdams, executive editor of *Waste Age*, who was responsible for much of the original editing of the pieces; to Chaz Miller, recycling manager for the Environmental Industry Associations, whose contributions form a substantial part of this book; all of the *Waste Age* writers and editors represented in this book and all those who contributed to the topic but for whatever reason are not represented here; to Karen Finkel and Kim Leaird, who created the original design and format for these articles as they appeared in *Waste Age,* and the entire *Waste Age* production staff; to Carolyn Baird, *Waste Age* editorial assistant, who painstakingly proofed the manuscript; to Eugene J. Wingerter, executive director and CEO of the National Solid Waste Management Association (NSWMA) from 1972 to 1994; and to Allen "Mike" Frischkorn, current president of the Environmental Industry Associations, the reconstituted umbrella organization for NSWMA, the Waste Equipment Technology Association, and the Hazardous Waste Management Association.

The greatest thanks is given to the people in the solid waste industry who have, through a long process, developed recycling in order to meet their customers' needs and keep the earth whole.

John T. Aquino
Editor-in-Chief/Publishing Director
Waste Age Publications
Washington, D.C.

John T. Aquino has been editor-in-chief/publishing director for *Waste Age* Publications since January 1991. He was previously vice president of Hanley-Wood, Inc., publisher of *Builder* and *Remodeling* magazines, and editor of *Mortgage Banker* magazine and *Music Educators Journal.* He is the author of seven collections/bibliographies for associations, including *Performance-Based Teacher Education: An Annotated Bibliography* and *Careers in Music*, and five monographs and numerous articles on his own. He was born in Washington, D.C. and lives in Silver Spring, Maryland, with his wife Deborah.

***Waste Age* Publications staff.**

Waste Age is the leading publication in the solid waste industry. It is owned by the Environmental Industry Associations and is published monthly. In 1993, it was selected as the best publication in the field of manufacturing by the editors of *Magazine Week* and *Folio*.

Recycling Times is *Waste Age*'s biweekly sister publication, covering recycling markets.

The Environmental Industry Associations—the National Solid Wastes Management Associations, the Hazardous Waste Management Association, and the Waste Equipment Technology Association—is the leading association in the solid waste industry.

CONTRIBUTORS

John T. Aquino
Editor-in-Chief/Publishing Director
Waste Age Publications
Washington, D.C.

Richard B. Curtis
Manager, Equipment and Safety
Waste Equipment Technology Association
Washington, D.C.

Jennifer A. Goff
Managing Editor
Recycling Times
Washington, D.C.

Mike Holderness
Writer
London, England

Michael Knoll
Project Coordinator
Department of Sanitation
New York, New York

John A. Legler
Executive Vice President
Waste Equipment Technology Association
Washington, D.C.

Michael G. Malloy
Contributing Editor, *Waste Age*
Editor, *Infectious Waste News*
Washington, D.C.

Chaz Miller
Recycling Manager
Environmental Industry Associations
Washington, D.C.

Michael Misner
Former Market Analyst
Recycling Times
Washington, D.C.

Robert W. Ollis, Jr.
Partner
Chapman and Cutler
Chicago, IL

Tom Polk
Project Manager for Recycling
Industries Development
Division of Business Development
Maryland Department of
Economic and Employment
Development
Baltimore, Maryland

Lisa Rabasca
Editor
Recycling Times
Washington, D.C.

Adrienne Redd
Former Assistant Recycling
Coordinator
Bethlehem, Pennsylvania
Freelance Writer
Lansdale, Pennsylvania

Christina Thoresen
Former Editorial Assistant
Waste Age Publications
Washington, D.C.

Kathleen M. White
Senior Editor
Waste Age Publications
Washington, D.C.

Randy Woods
Managing Editor, *Waste Age*
Associate Editor, *Recycling Times*
Washington, D.C.

LIST OF TABLES

CONTENTS

Waste Age/ Recycling Times'

RECYCLING HANDBOOK

1 WASTE PRODUCT PROFILES

By *Chaz Miller*, recycling manager for the Environmental Industry Associations, Washington, D.C.

What are the components of the waste stream? What are their characteristics? What are the markets for these materials as recyclables? This chapter is primarily composed of a series of profiles by Chaz Miller, recycling manager for the Environmental Industry Associations, Washington, D.C. These profiles are brief, factual listings of the solid waste management characteristics of materials in the waste stream. Each profile highlights a product, discusses how it fits into integrated waste management systems, and provides current data on recycling and markets for the product. These profiles were originally written in 1993 and 1994. They have been updated by the author for this book, using the latest price information and 1993 data published in late 1994 by the U.S. Environmental Protection Agency in Characterization of Municipal Solid Waste in the United States: 1994 Update. *The chapter concludes with an article on the issue of "quality" recyclables, which, although it includes interviews with individuals that were conducted in 1991, provides advice that is still useful today.*

GLASS CONTAINERS

Glass Container Solid Waste Facts

Weight: In 1993, glass containers constituted 12.2 million tons or 5.9% of the municipal solid waste (MSW) stream before recycling. The average weight of a glass bottle is one-half pound.

Volume: Glass containers comprised 1.5% of landfilled MSW by volume in 1993. Whole glass containers have a density of 156 pounds per cubic yard. Crushed glass can have a density of up to 2,700 pounds per cubic yard.

Recycling Rate: 20-25% (1991).

Recycled Content: 30% (includes in-house scrap, 1991).

Value: Glass containers have a low per-ton value. Currently, glass container plants will pay $40–$55 per ton (clear bottles), $20–$55 (brown bottles), and $0–$20 (green bottles), while processors generally only pay $10–$20 per ton for clear or brown and $0 for green. Nonglass container markets generally pay less than $5 per ton.

Amount Available for Recycling: In 1990, U.S. production was 10.3 million tons or 41.1 billion containers. The actual amount of imported containers is unknown, although estimates range from 750,000 tons to 1.25 million tons.

Food containers are 33% of U.S. production; beer bottles are 31%; beverage containers are 22%; wine and liquor bottles are 9%; the remainder is cosmetic and pharmaceutical bottles. Imported containers are primarily beer, wine, and liquor bottles.

Two-thirds of American-produced bottles are "flint" (clear), one quarter are "amber" (brown), and the remainder are various shades of green and a very small amount of blue bottles. At least half of imported bottles are green.

Glass Containers and Integrated Waste Management

Source Reduction: Source reduction of glass containers is primarily achieved through refillable bottles. However, American consumer preference for "convenience" has led to a precipitous decline in the production of refillable containers in the last 20 years. Today, the major market for refillables is beer consumed on premises in bars and restaurants. (Note: lightweighting is often considered source reduction. Glass container manufacturers, like all package manufacturers, constantly seek to use less material to make bottles. However, volume reduction is true source reduction for solid waste management.)

Recycling: Glass containers are highly recyclable and can be made back into glass containers with little loss of material.

Composting: Glass is nondegradable. MSW compost operations will attempt to exclude glass by hand-picking or mechanical means, or will grind it into a grit-like substance.

Incineration: Glass is noncombustible and generally forms a slag on incinerator equipment. In addition, the abrasiveness of glass causes problems with grates and conveyor equipment.

Landfilling: Glass is nonbiodegradable and chemically inert.

Glass Recycling Past and Present

The glass container industry has always used in-house scrap ("cullet") as raw material because it melts at a lower temperature than virgin raw materials. When improvements in the production process caused a decline in the amount of in-house cullet, manufacturers started using post-consumer cullet as a raw material. The first organized attempt to collect post-consumer glass bottles occurred in Bridgewater, NJ, in 1968.

Beverage container deposit legislation, originally intended to reduce litter, spurred glass recycling by providing a stable supply of high-quality cullet. Because deposits caused a loss of market share, the glass container industry promoted curbside collection systems as the most efficient way to collect all types of glass containers. This effort paid off in 1990 when New Jersey announced a glass recycling rate of 53%, the highest in the nation. In addition to curbside programs, bars and restaurants are also prime areas for glass recycling programs.

Markets for Glass Containers

The primary market for glass containers is the 73 glass container manufacturing plants in the United States. Other markets include road construction, either on the surface ("glasphalt") or as a roadbase aggregate; filler in storm drain and French drain systems; fiberglass production; abrasives; glass foam; and glass beads for reflective paint.

Limitations to Glass Recycling

Very Strict Raw Material Specifications: Contaminants in glass container production include:

- noncontainer glass which is chemically different from container glass;

- mixed-color glass (melting mixed-color glass produces off-color bottles and can lead to a foaming reaction in a furnace);
- ceramics such as coffee cups and swing-top closures found on several imported beers; and
- heat-resistant glass (both ceramics and heat resistant glass do not melt at the temperatures used in a glass container furnace and show up in a bottle as a "stone" or other defect).

Color Separation: Efficient mechanical systems for color separation are currently nonexistent. As a result, glass must be color-separated by hand. This is expensive and time-consuming.

Other limitations to glass recycling include a glut of green containers, due to large numbers of imported green bottles and limited American production of green bottles; potentially high transportation costs; broken glass contamination of other recyclables in some collection systems; and the very low costs for virgin raw materials such as sand and limestone.

Sources

Characterization of Municipal Solid Waste in the United States: 1990 Update, U.S. Environmental Protection Agency.

Current Industrial Report - Glass Containers, Series M32G - Final 1990 Report (Bureau of Census, Department of Commerce).

Glass Packaging Institute.

New Jersey Office of Recycling, 1990 Report.

NSWMA Technical Bulletin 85-6.

Waste Age's Recycling Times, March 10, 1992.

Resource Recycling Technologies.

U.S. Industrial Outlook 1991 - Cans and Containers (Department of Commerce).

****

NEWSPAPERS

Newspaper Solid Waste Facts

(NOTE: Newsprint is the paper itself. Newspapers are what newsprint becomes after it is printed on.)

Weight: In 1993, newspapers constituted 12.9 million tons, or 5.9% of the municipal solid waste (MSW) stream before recycling.

Volume: In 1993, newspapers comprised 4.0% of the volume of landfilled MSW.

Recycling Rate: In 1991, 52% of the newspapers consumed in the U.S. were recycled. In 1988, 33% were recycled.

Recycled Content: Newsprint can be composed of 0% to 100% recycled fiber. In 1991, recycled newsprint consumed 11.5% of the newspapers sold in the U.S.

Value: Newspapers have gone from a low to negative per-ton value in the early 1990s when, in many parts of the country, processors had to be paid to take newspapers. In 1994, old newspaper (ONP) prices increased drastically, with processors generally paying $10–30/ton. Mill prices for baled, curbside-collected newspaper range from $60–$100 per ton (prices as of December 1993).

Amount Available for Recycling: U.S. newsprint production capacity was 7 million tons in 1991, with total North American capacity of 18.5 million tons. Due to overexpansion and the effects of the recession on newspaper sales, the newsprint industry currently has excess production capacity. Over 50% of the newsprint consumed in the U.S. is imported from Canada.

In 1993, 7 million tons of newspaper were collected but went unrecycled. This is 4.3% of MSW after recycling.

Newspapers and Integrated Solid Waste Management

Source Reduction: Source reduction is hard to achieve for newspapers. Paper can be lightweighted and downsized, but significant additional lightweighting seems unlikely. Newsprint production decreases primarily when an economic recession causes fewer ads to be printed, resulting in fewer newspaper pages. As a result of the recession, newsprint consumption for 1991 was down by 600,000 tons compared to 1990 figures.

Recycling: Newspapers are highly recyclable and can be manufactured back into newsprint, using the two most common methods of deinking: flotation and washing. However, because the recycling process shortens paper fibers, a newspaper can be recycled a maximum of six to eight times before the fibers are too short to be used. Virgin pulp will always be necessary for newsprint production.

Composting: Newspaper is organic and highly compostable with only trace amounts of ink in the compost.

Incineration: Newspaper is easily combustible. A pound of newspaper has 7,500 Btus (compared to 4,500–5,000 Btus for an average pound of MSW).

Landfilling: Newspapers, like most materials, degrade very slowly in a modern landfill.

Current Newspaper Recycling Efforts

Traditionally, newspapers are the most recycled product in America. For decades, volunteer organizations raised funds by collecting newspapers when market demand was high. In 1970, the response to the first Earth Day led to an increased emphasis on recycling. Many municipalities, primarily in the Northeast, began curbside collection of newspapers. The number of curbside programs went from 47 in 1974 to 240 in 1980. Today, over 3,000 communities collect newspapers at the curbside for recycling.

Markets for Newspapers

The largest market for newspapers is the recycled newsprint industry, which recovers clean pulp from old newspapers through a process known as deinking. The first successful newspaper deinking mill was founded in 1953 by the Garden State Paper Company headquarters. By 1988, seven deinking mills, with a combined capacity of 1.5 million tons were operating in North America. However, the surge of curbside collection programs in the late 1980s overwhelmed the market for newspapers, creating a supply glut, negative prices, and occasional instances of newspapers being landfilled after they were collected for recycling.

As a result, Connecticut and California passed legislation requiring that newspapers sold in those states contain prescribed amounts of newsprint produced from post-consumer newspaper. These laws were followed by similar laws in nine other states (Arizona, Washington

D.C., Illinois, Maryland, Missouri, North Carolina, Oregon, Rhode Island, Texas, and Wisconsin) and voluntary agreements in an additional 13 states. Faced with the prospect of a guaranteed demand for deinked newspaper, deinking capacity has increased with the construction of new mills and expansions at existing mills (including mills that previously only used virgin fiber). Current estimates project that by 1995, 9.3 million tons of old newspaper will be consumed by North American deinking facilities.

Other markets for newspapers include boxboard and other recycled paper products, exports (primarily to Pacific Rim mills using wastepaper as a primary raw material), cellulose insulation, and animal bedding.

Grades of Newspaper for Recycling

Processors normally buy loose (unbaled) newspapers. They then sort and pull out contaminants, bale the newspapers and sell them to an end market. Cleanup, storage, baling, and transportation costs can run from $25–$30 per ton for a processor. Baled newspaper can be sold as one of a number of different grades. Number 6 news is the grade most commonly used for curbside-collected newspapers. The specifications for Number 6 news allow for no more than 5% of "other papers," 0.5% prohibitive materials, and 2% total out-throws. Higher-numbered grades have tighter requirements, with Number 9 (over-issue news, which is not a post-consumer grade) being the highest.

Limitations to Newspaper Recycling

In addition to market capacity, the physical nature of groundwood fibers limits newspaper recycling. In the groundwood process, lignin remains in the paper fiber. Lignin causes newspaper to turn yellow and deteriorate after exposure to the elements or prolonged storage. As a result, supplies must be sold to an end market within six months (if they are stored properly).

Contamination during collection can be a major problem. Newspaper that gets wet or is mixed with food, broken glass, or other grades of paper will be hard to sell. In addition, the export market—which is the third largest market for old newspapers—can be subject to highly volatile price and buying swings caused by local and international economic effects.

Sources

American Newspaper Publishing Association (ANPA, Springfield, VA).
American Paper Institute (API, New York, NY).
Characterization of Municipal Solid Waste in the United States: 1990 Update, U.S. Environmental Protection Agency.
Facing America's Trash, Office of Technology Assessment, 1989.
Waste Age's Recycling Times, March 10, 1992.
Scrap Specifications Circular 1991, Institute of Scrap Recycling Industries (ISRI, Washington, DC).

****

POLYETHYLENE TEREPHTHALATE (PET)

Note: Polyethylene terephthalate (PET aka PETE) is a plastic resin used primarily to make soft drink bottles. Peanut butter, salad dressing, and other household and consumer products are also packaged in PET bottles. Other forms of PET packaging include trays and sheeting for cups and food trays.

PET Solid Waste Facts

Weight: In 1993, PET soft drink bottles—which constituted two-thirds of PET packaging in 1993—contributed 560,000 tons, or 0.3% of the municipal solid waste (MSW) stream before recycling.

Volume: PET soft drink bottles comprised 0.4% of landfilled MSW by volume in 1993. Whole PET bottles have a density of 30–40 pounds per cubic yard; granulated PET bottles can have a density of 700–750 pounds in a corrugated shipping container; and landfilled PET bottles have a density of 355 pounds per cubic yard.

Recycling Rate: The PET bottle recycling rate in 1993 was 8.6%.

Recycled Content: Most PET bottles have no recycled content. However, some soft drink companies use either the methanolysis or glycolysis process to produce PET soft drink bottles with up to 25% recycled content. These processes depolymerize the PET resin into its original two components, dimethyl terephthalate and ethylene glycol, and then use these two components to produce new PET. Some ketchup bottles use recycled PET in a middle layer of the bottle, while some household product bottles can use up to 100% recycled PET. In this case, the PET resin is not depolymerized, but is simply cleaned, reground, and remolded.

Value: PET has a relatively high per-ton value, but a low per-container value. Currently, processors pay on the range of 0–5 cents per pound for clear containers (unbaled), while end markets pay from $40 to $160 per ton. (It takes 7.5 2-liter PET soft drink bottles—by far the most common PET bottle—to make a pound of PET bottles and 15,000 to make a ton.) Mixed color PET has a lower value, while granulated PET has a higher value.

Amount Available for Recycling: In 1991, 686,000 tons of PET packaging were consumed in the U.S. Of that, 396,500 tons were soft drink bottles, 203,500 tons were custom bottles, and 86,000 tons were other forms of PET packaging.

Most PET bottles are clear, with brown or green bottles also being used.

The average American uses almost five pounds of PET packaging each year.

PET constituted 9% of the plastic packaging and 28% of the plastic bottles sold in the U.S. in 1990.

PET Packaging and Integrated Waste Management

Source Reduction: The PET container allows greater amounts of product to be packaged in a lighter-weight bottle than alternative containers. In addition, the PET container is 28% lighter now than 20 years ago. However, PET's low weight is offset by its high volume, with a volume-to-weight ratio of 2.5 to 1.0.

Recycling: PET is a highly recyclable plastic that can be made into carpet fiber, industrial strapping, fiberfill for sleeping bags and ski jackets, and into many other products. Recent breakthroughs in resin regeneration also allow PET bottles to be made back into PET bottles.

Composting: PET is inorganic. Composting operations will attempt to exclude PET by handpicking, or will grind it into a grit-like substance that will not biodegrade.

Incineration: PET is highly combustible, with a per-pound Btu value of 10,933 (compared with 4,500 Btus for a pound of MSW).

Landfilling: PET is nonbiodegradable.

PET Recycling Past and Present

PET is a young plastic resin, created in 1973. While industrial PET scrap has always been reused, post-consumer PET recycling only began in the early 1980s. Beverage container deposit legislation, originally intended to reduce litter, provided a stable supply of easily recyclable PET bottles (although Iowa had to ban PET deposit containers from its landfills to ensure recycling). Currently, the great majority of recycled PET bottles are soft drink containers from deposit and nondeposit states. As a result, PET soft drink bottles are the most widely recycled plastic package. However, curbside collection programs are increasingly expanding their PET collections to include all PET bottles.

The resin code for PET is "1." This code can be found on the bottom of virtually all PET containers.

Market for PET Containers

The primary market for recycled PET is the fiber industry, which uses PET for carpet fiber and other products. Other markets include "closed-loop" recycling back into PET containers, and various mixed plastic recycling markets, such as plastic lumber.

Limitations to PET Recycling

Strict Raw Material Specifications: Contaminants in PET recycling include other plastics, particularly polyvinyl chloride (PVC), which has the same specific gravity as PET, yet is highly incompatible with most PET recycling uses. PVC usually contaminates PET shipments through inadvertent commingling of PVC and PET bottles (PET and PVC bottles look alike) or through inclusion of bottle caps with PVC liners.

A mechanical system for separating PVC from mixed plastic recyclables has been successfully tested and is in operation in a number of recovery facilities.

Color Separation: Generally, green or brown PET bottles must be kept separate from clear PET bottles because markets for mixed colored PET are limited, and pigmented PET has a slightly different chemistry than clear PET.

Collection: PET's high volume-to-weight ratio creates problems in collecting PET in curbside collection programs. On-board densification of PET reduces the volume of the containers, but adds an extra time element for on-route processing. Other options include wire cages or other special compartments on a truck dedicated to plastic containers.

Transportation: PET must be baled in order to achieve transportation efficiencies. Granulating improves efficiencies, but markets for granulated PET are limited, due to concerns about potential contamination.

Sources

Characterization of Municipal Solid Waste in the United States, 1990 Update, U.S. Environmental Protection Agency.
Modern Plastics.
National Association for Plastic Container Recovery.
National Recycling Coalition Measurement Standards and Reporting Guidelines.
Post-Consumer Plastics Recycling Rate Study, Council for Solid Waste Solutions.
Waste Age's Recycling Times, April 21, 1992.
Trash to Cash, Investor Responsibility Research Center, 1991.

****

ALUMINUM PACKAGING

Aluminum Solid Waste Facts

Weight in MSW: In 1993, aluminum packaging constituted 2.0 million tons of municipal solid waste (MSW) before recycling. Of this, 1.6 million tons were soft drink, beer, and food cans. Aluminum cans account for more than 95% of used beverage cans. However, aluminum cans only have 5% to 10% of the food can market.

The average weight of an aluminum can is 0.554 ounces (28.87 cans per pound).

Foil packaging amounted to 300,000 tons of MSW. Foil packaging can be wrapping foils, semi-rigid packaging (pie plates, frozen food trays), and flexible packaging (cigarette foil and candy wrappers).

Volume in MSW: Aluminum packaging made up 1.4% of landfilled MSW by volume in 1993. Whole aluminum cans have a density of 50–74 pounds per cubic yard, and landfilled cans have a density of 250 pounds per cubic yard. Landfilled foil has a density of 550 pounds per cubic yard.

Recycling Rate: In 1993, 63.4% of aluminum cans used in the U.S. were recycled (this equals 56.845 billion cans or one million tons of aluminum). The foil recycling rate is unknown, but is probably no more than 1% or 2%.

Recycled Content: 52% in 1989 for cans, 0% for foil.

Value: Aluminum cans are one of the most valuable recyclables on a per ton basis. "Street" (buyback center or processor) prices range from 20 to 32 cents per pound ($400–$640 per ton). End user (toll) prices range from 35 to 44 cents per pound ($700 to $880 per ton). Prices for aluminum foil range from 5 to 15 cents per pound.

Amount Available for Recycling: In 1991, 1,576,923 tons of aluminum cans were produced and used in the U.S. Only a minimal amount of aluminum cans are imported.

In 1991, 133,000 tons of aluminum wrapping foil, 66,500 tons of aluminum semi-rigid containers, and 100,000 tons of flexible packaging were used in the U.S.

The average American uses 364 aluminum cans and 2.4 pounds of aluminum foil per year.

Aluminum Packaging and Integrated Waste Management

Source Reduction: Aluminum cans are a low-weight, relatively high-volume container with a 2.75:1 volume-to-weight ratio. In the last 20 years, aluminum containers have been downsized by 30%. A high recycling rate combined with source reduction attributes make aluminum cans a very "green" container.

Foil is also a low-weight, highly source-reduced form of packaging, with a 1.5:1 volume to weight ratio.

Recycling: Aluminum containers are highly recyclable and can be made back into aluminum containers, with melt loss of only 5% to 15%. Aluminum foil recycling is in its infancy.

Composting: Aluminum is nondegradable. Composting operations will attempt to exclude aluminum by mechanical means or by handpicking.

Incineration: Aluminum is noncombustible and can end up as a residue in combustion ash.

Landfilling: Aluminum is nondegradable.

Aluminum Recycling Past and Present

Industrial aluminum scrap has been recycled for as long as aluminum has been used in the manufacturing industry. The aluminum can, however, is a relatively young container, first appearing commercially in 1953. Aluminum-can recycling began in earnest in the late 1960s. Offering a cash value for empty cans, aluminum companies quickly discovered their product was both highly and easily recyclable. Whether using mobile or permanent collection centers, or encouraging fund-raising collections by volunteer groups, the aluminum recycling rate has grown steadily over the last two decades.

Aluminum recycling is a relatively simple operation: recovered cans are shredded, melted, cast into ingot, rolled into can sheet, and transformed back into cans. It can take as little as six weeks for an aluminum can to be manufactured, used, and recycled back into an aluminum can.

Markets for Aluminum Packaging

The primary market for aluminum containers is made up by the aluminum producing companies that buy back used beverage containers and use them to make new can sheet. Secondary aluminum smelters that manufacture aluminum ingot from scrap, however, are an increasing market for aluminum containers. Secondary markets often function as "tollers" of scrap metal for integrated companies.

Limitations to Aluminum Can Recycling

Aluminum cans can be contaminated by a number of items, including dirt, moisture (concentrated moisture in a furnace can cause a steam explosion), plastic (which can burn in delacquering operations), glass, noncontainer aluminum, and other metals. Lead is a particular problem because dishonest suppliers will put lead sinkers and lead tire weights into empty cans to gain extra weight. While a magnet will easily separate aluminum cans from steel cans, other contaminants require vigilance on the part of aluminum can recyclers.

Limitations to Foil Recycling

One of the new kids on the block, foil recycling is limited by the need to clean used foil packaging of food debris and by the small amounts of foil packaging generated by individuals.

Foil uses a different alloy than cans. As a result, foil can be a contaminant in cans, and cans can be a contaminant in foil.

Sources

Aluminum Recycling: Vital to You; Vital to Us, Aluminum Association, 1991.
Characterization of Municipal Solid Waste in the United States: 1992 Update, U.S. Environmental Protection Agency, Office of Solid Waste, 1992.
Measurement Standards and Reporting Guidelines, National Recycling Coalition, 1989.
Waste Age's Recycling Times, July 14, 1992.
Rigid Container Recycling, Department of Commerce, 1991.
Trash to Cash, Investor Responsibility Research Center, Washington, D.C., 1991.

****

HIGH-DENSITY POLYETHYLENE (HDPE) BOTTLES AND CONTAINERS

NOTE: HDPE packages are made from resin produced from the chemical compound ethylene. The resin is either blow-molded to make bottles or injection-molded to make containers. HDPE resin is naturally milky white and is commonly used for milk, water, and juice bottles. Colorants can be added to the resin and used to make bottles for household and consumer products such as detergents, shampoos, and motor oil. HDPE containers are used for dairy products and other consumer products. Base cups for soft drink bottles are often HDPE. Other HDPE packaging includes caps and films for bags and sacks. This profile only covers HDPE bottles and containers.

HDPE Solid Waste Facts

Weight in MSW: In 1993, HDPE milk bottles made up 0.3% (553,000 tons) of municipal solid waste (MSW) before recycling. All HDPE bottles and containers made up approximately 0.6% (1.2 million tons) of MSW.

Volume in MSW: In 1993, milk bottles made up 0.5% of landfilled MSW by volume. All HDPE bottles and containers made up 1.8% of landfilled MSW by volume. Milk bottles have a landfill density of 355 pounds per cubic yard.

Milk bottles have a density of 30 pounds per cubic yard (whole) and 65 pounds per cubic yard (flattened). Colored HDPE bottles have a density of 45 pounds per cubic yard (whole) and 65 pounds per cubic yard (flattened). Granulated HDPE will have a "gaylord" box density of 800–1,000 pounds. Bales of HDPE generally weigh 500–800 pounds.

Recycling Rate: In 1991, approximately 135,000 tons of HDPE bottles and containers were recycled in the U.S., a 23.6% recycling rate.

Recycled Content: In spite of attempts to increase the percentage of "bottle-to-bottle" HDPE recycling, the use of recycled content is still limited in packaging. While some HDPE consumer product bottles use 50% post-consumer recycled content, 25% is more common for those with recycled content.

Value: End-users generally pay $100–140 per ton for milk bottles, while processors generally pay zero to $20 per ton. Mixed color HDPE has a lower value, with most processors paying $60–80 per ton (although the East Central region has end markets paying $120–140 per ton). Mixed HDPE/PET loads generally receive zero to $40 per ton from processors.

Amount Available for Recycling: In 1991, 484,000 tons of milk and water bottles; 454,000 tons of household chemical products bottles; 102,000 tons of motor oil bottles; and 109,000 tons of cosmetic and pharmaceutical bottle resin were sold in the U.S., supplying 1.159 million tons of HDPE packaging. More than half are colored. In addition, 208,000 tons of injection-mold HDPE container resin were sold in the U.S. In all, 1.367 million tons of HDPE bottles and containers were sold in 1991.

The average American uses nearly 12 pounds of HDPE bottles and containers in one year.

HDPE constitutes 21% of the plastic in MSW and 50–60% of plastic bottles.

HDPE Bottles and Containers and Integrated Waste Management

Source Reduction: HDPE bottles and containers are lighter-weight than most alternative, nonplastic packages. As with all packages, they have successfully downsized over the years. In the 1970s, empty milk jugs weighed about 100 grams. Now, they weigh 60 grams. However, HDPE bottles and containers have a negative volume-to-weight ratio of 3:1 (their volume percentage in MSW is three times as great as the weight percentage).

Recycling: HDPE can be made into a wide variety of products including plastic lumber, recycling bins, grocery bags, mud flaps, flower pots, and inner layers for new plastic bottles.

Composting: HDPE is inorganic. Composting operations will attempt to exclude HDPE packaging by handpicking or mechanical means such as trommels.

Incineration: HDPE is highly combustible, with a per-pound Btu value of 18,690 Btus (compared with 4,500–5,000 Btus for a pound of MSW). Ironically, this very high Btu level may cause problems in boilers with low per-pound Btu ratings.

Landfilling: HDPE is nonbiodegradable.

HDPE Recycling Past and Present

Recycling of industrial HDPE scrap is not new. Recycling of post-consumer HDPE began with HDPE base-cups from PET bottles collected as a result of container deposit laws. Curbside collection programs added milk bottles because they are easy to identify and separate. Other HDPE bottles are now being collected, but containers remain largely uncollected.

The recycling code "2" is used to distinguish HDPE. However, this does not distinguish between blow-molded and injection-molded HDPE packages.

Markets for HDPE

After collection (and processing of commingled plastics at a materials recovery facility, MRF), most HDPE is sold to plastics processors. While procedures vary, the plastic is separated from bottle components (caps, labels, and their adhesives), washed, dried, and ground into flakes. Some processors produce pellets from the flakes.

Limitations on HDPE Recycling

Value: Currently, post-consumer HDPE resin is more expensive than virgin resin, due to the cost of collecting and processing HDPE bottles and to overcapacity in virgin resin production, which has caused low prices for virgin resin. In addition, many nonbottle markets are low-value markets.

Color Separation: In the recycling process, mixing colored HDPE bottles will contaminate natural bottles, leading to a dark, often black, product. Automatic systems for color separation are not commercially available.

Blow Mold Bottles and Injection Mold Containers

HDPE bottles are manufactured using a blow-molding technique, while tubs, base-cups, and other rigid HDPE packaging are produced by injection-molding. Blow-molding and injection-molding have a different melt flow index. As a result, the two types of HDPE can be incompatible in reprocessing operations. For instance, mixing blow-molded bottles with injection-mold containers to make HDPE pipe will cause stress fractures in the pipe.

Collection: HDPE's high volume-to-weight ratio creates collection problems in curbside programs. Large milk and detergent containers pose a particularly vexing problem. On-board densification systems reduce the volume problem, but can add to route time. Other solutions include wire cages dedicated to HDPE.

Sources

Characterization of Municipal Solid Wastes in the United States: 1990 Update, U.S. Environmental Protection Agency, June, 1990.
Design for Recycling: A Plastic Bottle Recycler's Perspective, Society of the Plastics Industry (SPI), February, 1992.
Modern Plastics, January, 1992.
National Recycling Coalition Measurement Standards and Reporting Guidelines, NRC, Washington, D.C., October, 1989.
Plastic Packaging Opportunities and Challenges, Testin and Vergano, SPI, 1992.
Partnership for Plastics Progress, Fact Sheet/News Release, June 1, 1991.
Post Consumer Plastic Recycling Rate Study, Council for Solid Waste Solutions, Washington, D.C., November, 1991.
Waste Age's Recycling Times, June 16, 1991.
Resource Recycling Technologies, Vestal, NY.
The Solid Waste Management Problem, Council for Solid Waste Solutions, Washington, D.C., 1989.
Trash to Cash, Investor Responsibility Research Center, Washington, D.C., 1991.

****

CORRUGATED BOXES

Corrugated boxes (also known as old corrugated containers, or OCC) are used to ship products to factories, warehouses, retail stores, offices, and homes. Corrugated boxes usually have a fluted corrugated medium sandwiched between layers of linerboard. The term "corrugated" also refers to cuttings generated from the manufacturing of corrugated containers. Corrugated boxes are often mistakenly referred to as "cardboard" boxes.

OCC Solid Waste Facts

Weight: In 1993, OCC constituted 26.3 million tons, or 12.7% of the municipal solid waste (MSW) stream before recycling.

Volume: In 1993, OCC comprised 7.2% of landfilled MSW by volume. Unbaled OCC has a density of 350 pounds per cubic yard, baled boxes have a density of 1,000 to 1,200 pounds per cubic yard. With a landfilled density of 750 pounds per cubic yard, OCC is exceeded only by yard waste in volume impact at MSW landfills.

Recycling Rate: In 1993, the OCC recycling rate reported by the U.S. EPA was 55% (14.6 million tons). In 1993, the OCC recycling rate reported by industry was 62% (16.7 million tons). The EPA rate only covers post-consumer corrugated waste. The industry-reported number includes converter scrap from the manufacturing of corrugated boxes out of linerboard and corrugated medium.

Recycled Content: Most corrugated containers have some recycled content, primarily in the corrugated medium, but also, to an increasing extent, in the inner and outer linerboard layers. In 1990, corrugated boxes had a recycled content level of about 21%.

Value: OCC value varies greatly both by location and by the manner of preparation. Some processors require a tipping fee for unbaled OCC in the Northeast, whereas in other sections of the U.S., processors will pay from $10 to $35 per ton for unbaled OCC. End markets also show regional variation, with prices as low as $10 per ton in the Northeast to as high as $59 per ton in the West.

The containerboard industry is in the midst of a major expansion in capacity to produce both linerboard and corrugated medium from OCC and other sources of secondary fiber. As a result, prices for OCC will improve as capacity expands.

Amount Available for Recycling: In 1993, 11.7 million tons of OCC were not collected for recycling.

Most corrugated boxes have a brown-colored outer layer. Some, however, use a white-colored outer layer, known as "mottled white." Several linerboard manufacturers are now using office waste paper in the outer white layer.

In 1990, 19 pounds of corrugated containers were produced for every American.

Corrugated containers constitute the largest paperboard grade used in the U.S. and are exceeded in the paper category only by the category of printing, writing, and related paper.

Corrugated containers contribute equally to MSW on a weight and volume basis with a 1:1 volume to weight ratio.

OCC and Integrated Waste Management

Source Reduction: As with all packages, corrugated boxes can be lightweighted. Using less linerboard has resulted in a 10–15% weight reduction in the last 10 years. The need to meet compression, stacking strength, and burst tests limits the ability to lightweight corrugated boxes. Ironically, heavy use of recycled content fibers can increase the weight of a corrugated container in order to meet the requirements of these tests.

Recycling: Corrugated boxes are easily and highly recyclable. OCC leads all materials in amount of material recycled. Only aluminum cans have a higher recycling rate than corrugated boxes.

Composting: Corrugated boxes, if shredded properly, are easily compostable.

Incineration: Corrugated material is easily combustible, with a per pound Btu value of 7,047 Btus (compared with 4,500 Btus for a pound of MSW).

Landfilling: OCC, like most materials, degrades very slowly in a modern landfill.

OCC Recycling, Past and Present

Old corrugated containers have been recycled by large producers such as grocery store warehouses and factories for some time. These outlets produced sufficient amounts of OCC to justify the cost of equipment to bale OCC in order to sell it to waste haulers or paperstock dealers. As the interest in recycling increased, so did the amount of OCC diverted for recycling. Today, the main sources of unrecycled corrugated boxes are residences and small retail outlets.

Market for Corrugated Containers

The primary market for OCC is the paper industry, which uses OCC for corrugated medium, linerboard, recycled paperboard, and other paper products. In addition, almost three million tons of OCC were exported in 1991.

Limitations to Corrugated Recycling

Strict Raw Material Specifications: Contaminants in OCC include wax coatings, plastics, chipboards, mill wrappers, food contamination, and garbage. Another contaminant is a corrugated box called "yellow corrugated." Composed of often-recycled fibers that have lost much of their fiber length and, as a result, much of their fiber strength, this grade of OCC is yellowish in color and weaker than other forms of corrugated.

OCC is defined by the Institute of Scrap Recycling Industries' (ISRI) Paper Stock Institute as "baled corrugated containers having liners of either test liner, jute, or kraft. Prohibitive materials may not exceed 1%, total out-throws may not exceed 5%."

Collection: OCC is easy to collect from large commercial sources, especially warehouses. However it is cumbersome to collect from residences because of its bulkiness, and because individual residences generate a relatively small amount of OCC. As a result, residential collection programs are rare, but growing.

Processing: OCC has relatively low processing costs when it is collected with other grades of paper because corrugated boxes are easy to identify. The NSWMA's Materials Recovery Facility (MRF) Cost Study [see Chapter 3] shows a $42.99-per-ton average cost to process OCC at MRFs designed to handle loads of commingled residential recyclables, with a range of $20.29 to $50.56. MRFs designed to process loads of commercial recyclables will achieve lower processing costs due to economies of scale.

Sources

American Paper Institute News Release, March 9, 1992.

Characterization of Municipal Solid Waste in the United States, 1992 Update, U.S. Environmental Protection Agency, Office of Solid Waste.

Final Report, McDonald's Corporation & Environmental Defense Fund Waste Reduction Task Force, April 1991.
Jane Erkenswick, Fiber Options, a Division of Recycling Options, Chicago, IL.
Post Consumer Material Densities, Resource Recycling Technologies, Vestal, NY, 1991.
National Recycling Coalition Measurement Standards and Reporting Guidelines, 1990.
The Solid Waste Management Problem, Council for Solid Waste Solutions, 1990.
Waste Age's Recycling Times, September 8, 1992.
Trash to Cash, Investor Responsibility Research Center, 1991.

****

STEEL CANS

The steel food can was invented in England in the early 1800s. It is often called a tin can because of the thin layer of tin applied to the can's inner and outer surfaces. Tin is used to protect food and beverage flavors and to prevent rusting. Steel cans are made from tinplate steel, which is produced in basic oxygen furnaces. The steel beer can was first produced in 1938. A steel beverage can is often called a bimetal can because it uses a steel body and an aluminum top. Due to changes in steel making, an increasing amount—though not a majority—of steel cans are tin-free. These cans use a chromium wash to achieve the same results as tin.

Weight in MSW: More than 37 billion steel cans, or 2,900,000, were used in the U.S. in 1993. The great majority of steel cans are used for food products (steel cans account for more than 90% of food cans), followed by "general packaging" (paint cans, aerosol cans, etc.), then beer and soft drink cans, which account for less than 5% of beverage cans.

Steel cans constitute 1.0% of discarded municipal solid waste (MSW) by weight. The average weight of a steel can is 2.5 ounces.

Volume in MSW: Steel cans made up 1.3% of landfilled MSW by volume in 1993. Whole steel cans have a density of 150 pounds per cubic yard, baled cans have a density of 850 pounds per cubic yard, and landfilled cans have a density of 560 pounds per cubic yard.

Recycling Rate: 24.6% of the steel cans used in the U.S. were recycled in 1990, 46% were recycled in 1993 (this equals 1,380,000 tons of steel).

Recycled Content: Tinplate is produced in basic oxygen furnaces. In these furnaces, 20–30% of the raw material is normally scrap (including most, but not all grades of scrap). Steel can sheet manufacturers currently use slightly more pre-consumer than post-consumer scrap.

Value: Steel can value is steady, but not spectacular. "Street" (buyback center or processor) prices for unbaled cans range from $0 to $40 per metric ton (0–2 cents per pound). End-user prices for baled cans range from $32 to $87 per ton. (Prices as of November 3, 1992.)

Amount Available for Recycling: In 1993, 1.6 million tons of steel cans were landfilled.

The average American uses 142 steel cans (22.75 pounds) per year.

Steel Cans and Integrated Waste Management

Source Reduction: As with all containers, steel cans have downsized over the years. The bi-metal can is 40% lighter now than in 1970. The amount of tin used in a steel can has gone down dramatically from pre-World War II levels of 50 pounds per ton of tinplate steel to a current average of six pounds per ton.

Recycling: Steel cans are easy to recycle due to their magnetic properties and the limited number of potential contaminants in the remanufacturing process. Steel can be successfully separated from mixed solid waste, but will be dirtier than steel cans collected in a curbside collection program.

Composting: Steel cans are a contaminant in a composting program. However, they are biodegradable in the sense that steel cans left exposed to the elements will slowly rust. (Ironically, in the 1970s this was promoted as an environmental benefit of steel cans.)

Incineration: Steel cans are noncombustible and can end up as a residue in combustion ash. Waste-to-energy facilities commonly use magnets to remove steel cans before incineration. This steel, however, will be of lower quality than steel collected in curbside collection programs.

Landfilling: Steel cans will remain inert in a landfill.

Steel Can Recycling Past and Present

Industrial steel scrap has been recycled for as long as steel has been used in the manufacturing industry. The steel can, however, was not commonly recycled until the first post-consumer recycling programs in the mid 1970s. A sustained, nationwide effort to collect and recycle steel cans did not begin until 1988.

Steel can recycling is a relatively simple operation. First, cans are collected separately in curbside or other collection programs or separated from MSW by magnets. They are then shredded and either melted in a steel furnace or detinned and then melted.

Markets for Steel Cans

Steel can markets include the steel, detinning, and foundry industries. Steel mills vary in their ability to use steel can scrap. A basic oxygen furnace can only use scrap as 20–30% of its input, whereas an electric arc furnace can use 100% scrap. Electric arc furnaces, however, tend to have smaller capacities than basic oxygen furnaces. Detinners specialize in removing the tin from steel cans for resale in tin-using industries. Steel scrap from detinning is either sold to the steel industry or to the copper precipitation industry. Eight detinning plants are currently operating in the U.S. Continued decreases in the amount of tin used in steel cans has diminished the importance of this market. Iron and steel foundries are an emerging market for steel cans. Foundries use scrap as a raw material in making casts and molds for industrial users.

Limitations to Steel Can Recycling

Contaminants: Contaminants that can plague other materials are not a problem in a steel-making furnace. Aluminum tops from bi-metal cans provide a small heat boost to a furnace and the amount of tin in steel cans is now too little to create furnace problems.

Collection and Processing: The relatively low value and limited amount of steel cans in MSW have kept steel cans from being included in curbside collection programs in the past. However, many curbside and drop-off programs mix steel cans with aluminum cans to discourage theft of the aluminum cans.

Steel cans collected in a commingled collection program and processed at a materials recovery facility have an average processing cost of $67.53 per ton (with a median cost of $57.93 per ton).

Sources

Characterization of Municipal Solid Waste in the United States: 1992 Update, U.S. Environmental Protection Agency, Office of Solid Waste, 1992.

Measurement Standards and Reporting Guidelines, National Recycling Coalition, 1989.

Waste Age's Recycling Times, November 3, 1992.

"Post Consumer Material Densities," Resource Recycling Technologies, Vestal, NY.

Steel Can Recycling Institute, Pittsburgh, PA.

"Things You've Always Wanted to Know About Soft Drink Container Recycling," National Soft Drink Association, 1990.

Trash to Cash, Investor Responsibility Research Center.

U.S. Industrial Outlook, 1991—Cans and Containers, Department of Commerce, 1991.

****

OFFICE PAPER

Office paper (often called ledger paper) is a generic name given to a wide variety of paper products used in offices and businesses including letterhead, copying paper, file stock, and computer printouts. These paper grades usually have longer fibers and are brighter than newspaper and packaging grades. Letterhead and copying paper is usually "white," but can be produced in a wide variety of colors. Most office paper is made from chemically pulped paper fiber. However, some computer and writing paper is made from groundwood fiber (the same fiber used to make newspaper).

Office paper is a subcategory of the paper industry's "printing and writing" segment. Other paper products in this segment include book and magazine paper, junk mail, brochures, etc. Waste paper from offices can include these kinds of paper along with newspapers, corrugated containers, and other packaging grades of paper. This profile focuses on office "ledger" paper.

Office Paper Solid Waste Facts

Weight: According to the U.S. EPA, office paper constituted 7.1 million tons, or 3.4% of the municipal solid waste (MSW) stream before recycling in 1993. After recycling, office paper constituted 4.5 million tons, or 2.0% of MSW.

Using a broader definition of office paper, the National Office Paper Recycling Project (NOPRP), estimates 8.1 million tons of office paper were generated in 1990. NOPRP estimated an additional 2.8 million tons of newspaper and corrugated containers were in office paper waste streams in 1990.

Volume: In 1990, office paper comprised 2.6% of landfilled MSW by volume. Unbaled office paper has a density of 300–400 pounds per cubic yard; baled office paper has a density of 700–750 pounds per cubic yard; and landfilled office paper has a density of 800 pounds per cubic yard. Office paper contributes equally to MSW on a weight and volume basis with a 1:1 volume-to-weight ratio.

Amount Available for Recycling: In 1990, 51 pounds of office paper were produced for every American. Printing and writing paper comprises the largest segment (30%) of paper and paperboard products produced in the U.S.

Recycling Rate: In 1993, the office paper recycling rate reported by the EPA was 36.5% (2.6 million tons). The recycling rate in 1985 was 19.3% (600,000 tons). NOPRP estimated a 15.3% recycling rate in 1990 (1.3 million tons).

Recycled Content: The recycled content of office paper varies dramatically because of the diversity of office paper grades and the many differences between the manufacturers of those grades. While an overall recycled content percentage for office paper does not exist, the printing and writing paper segment had a 1990 waste paper utilization (recycled content) rate of 6%.

Value: Office paper is sold as a number of different waste paper grades. Cleaner, more homogenous grades (laser-free computer printout paper with no groundwood content) have a higher value than more heterogeneous grades (file stock). As with most paper commodities, high volumes of baled office papers receive higher prices. In many parts of the county, processors charge a tipping fee or pay nothing for mixed office paper; they pay $0–50 per ton for colored ledger; $10–100 per ton for white ledger; and $60–190 per ton for computer printout paper. End markets may charge a tipping fee for mixed office paper, although some end markets pay as much as $10 per ton (with higher prices in the Midwest). End markets for colored ledger pay from $25–80 per ton, white ledger from $75–185 per ton, and computer printout paper from $140–255 per ton (prices as of December 1992).

The printing and writing paper manufacturing industry is undergoing a major expansion in recycled content capacity, both in deinking capacity at paper mills and at a new generation of mills that just make deinked pulp. As a result, prices for all grades of office paper should improve as capacity expands.

Office Paper and Integrated Waste Management

Source Reduction: Source reduction is difficult to achieve for office paper. Ledger grades can be lightweighted and downsized, although additional lightweighting seems unlikely. Source reduction measures for office paper usually stress double-sided copying, switching from "legal-sized" paper to shorter paper, and using the lightest-basis weight stock possible for a job.

Recycling: Office paper is easily recyclable if the paper collected for recycling is well matched with the raw material specifications of its intended end user.

Composting: Office paper is relatively easy to compost, if shredded properly. Very low nitrogen content and lack of physical structure are inhibiting factors.

Incineration: Office paper is easily combustible with a per-pound Btu value of 7,200 Btus (compared with 4,500 Btus for a pound of MSW).

Landfilling: Office paper, like most materials, degrades very slowly in a modern landfill.

Office Paper Recycling Past and Present

Unprinted trimmings from printing and writing paper manufacturing operations and scrap from print shops have long been recycled because they are an accessible source of clean raw material. These grades, while they are the same type of paper as office paper, are considered pre-consumer paper, not post-consumer office paper. In addition, certain grades of post-consumer office paper from large generators, such as computer tab cards and computer printout paper from data processing centers, have traditionally been recycled.

Programs to recycle white ledger or all grades of office paper are more recent. In the mid-1970s the EPA promoted office paper recycling with the "Use It Again, Sam" program. In the '90s, as a result of commercial recycling mandates in a number of states and increased interest in recycling, office paper recycling programs are becoming commonplace.

Market for Office Paper

More than half of the office paper collected for recycling is exported, primarily to Pacific Rim countries. The next largest market is domestic tissue mills. Other uses for office paper include recycled boxboard, containerboard (including the outer liner in mottled linerboard), and printing and writing paper.

Currently, seven American paper mills deink office paper as the primary raw material in their production of printing and writing paper. These are small mills subject to economies of scale, resulting in paper that is generally more expensive than similar paper made from virgin pulp.

Recently, the number of mills dedicated to producing deinked market pulp from office paper, for sale to paper mills as a raw material, has grown dramatically. As a result, the Northeast and the Midwest are faced with the potential of an excess of capacity over local supply.

Limitations to Office Paper Recycling

Diversity of Paper Grades: Office wastepaper consists of many different types of paper, forcing recycling programs either to focus on a limited number of high-value grades (computer printout, white ledger, etc.) or to emphasize collecting a wide range of paper grades in which fiber types and paper colors are mixed together. In the latter case, the wastepaper will have a lower value because mixed wastepaper has low-value end markets or because of the cost of processing the paper to meet higher value specifications.

Strict Raw Material Specifications: High-value end markets for office paper have strict raw material specifications. They will exclude or strictly limit groundwood fibers, laser printing, colored paper, non-water-soluble glues, paper clips, rubber bands, and other items normally found in an office. Improved cleaning equipment and technological breakthroughs are minimizing the impact of some of these contaminants for some end-users.

Office paper is covered by at least six of ISRI's Paper Stock Institute wastepaper grades. In addition, individual mills have established office wastepaper grades. Recyclers must consult carefully with purchasers about specifications before delivering collected office paper to a market.

Collection: Mixed office paper is relatively easy to collect from offices. Essentially, all that is needed is to keep it separated from nonpaper waste throughout the generation and collection process. More valuable grades of office paper must be kept separate from nonpaper waste and from other paper grades. As a result, collection costs can be higher.

Processing: Clean grades of wastepaper (separated computer printout paper) are relatively inexpensive to process if the supplier has a reputation for delivering clean material. Mixed paper also can be inexpensive to process if it is sold as mixed paper. When supplies of higher grades of wastepaper are low, wastepaper dealers often can afford to pay processing costs to pull out higher value wastepaper from loads of mixed wastepaper.

Sources

Ron Albrecht, Ron Albrecht Associates, Annapolis, MD.

Characterization of Municipal Solid Waste in the United States, 1992 Update, U.S. Environmental Protection Agency, Office of Solid Waste.

Post Consumer Material Densities, Resource Recycling Technologies, Vestal, NY, 1991.

Recycled Papers: The Essential Guide, Claudia Thompson, 1992.

Waste Age's Recycling Times, December 17, 1992.

Resource Recycling, November, 1991.

Scrap Specifications Circular, 1991, Institute of Scrap Recycling Industries, Washington, D.C.

Supply of and Recycling Demand for Office Waste Paper, 1990 to 1995, National Office Paper Recycling Project, 1991.
Trash to Cash, Investor Responsibility Research Center, 1991.

****

YARD WASTE

Yard waste includes grass, leaves, and tree and brush trimmings. Grass is the biggest component of yard waste by weight (75%), followed by leaves (20%), and brush (5%). After paper and paper products, yard waste is the largest portion of the waste stream.

Weight in MSW: According to the U.S. EPA, in 1993 32.8 million tons of yard waste were generated in the U.S. This is 15.9% of municipal solid waste (MSW) by weight.

EPA yard waste weight data are based on sampling studies of mixed solid waste at transfer stations and landfills. Data are then extrapolated nationally. These studies are subject to error due to unusually wet or dry conditions, sampling mistakes, etc. In addition, EPA data do not include the amount of yard waste composted in backyards or grass clippings left on the lawn. However, use of the same methodology by the EPA over a 20–year period gives consistent baseline data for yard waste generation.

The average American "generates" 280 pounds of yard waste per year. Unlike most elements of the waste stream, yard waste generation varies dramatically based on a number of factors including climate, yard size, and percentage of population in multifamily housing. Yard waste can be the largest component of MSW during peak generation months in the summer and fall. During off-peak months, especially in winter, little yard waste is generated.

Composting Rate: The EPA estimated that in 1993, 6.5 million tons, or 19.8% of yard waste, was composted.

Amount Not Composted: In 1993, 26.3 million tons of yard waste were disposed of in landfills or waste-to-energy facilities. This is 16.2% of discarded (nonrecycled) MSW.

Volume in MSW: In 1993, the EPA estimated that yard waste occupied 35 million cubic yards of landfill space, or 8.1% of landfilled MSW by volume. Uncompacted yard waste has a density of 250–500 pounds per cubic yard; landfilled yard waste has a density of 1,500 pounds per cubic yard.

Value: Compost value is not tracked by national publications because of the seasonal nature of its production and differences in what is being sold, the end markets, and how the compost is sold.

Markets: High-quality compost can find a market as a soil amendment and as mulch for landscapers, farmers, nursery owners, and the general public. Local and state governments can use compost for highway verges, parks, and school grounds in place of top soil and mulches. Farm soil restoration is a potential high-growth market. In some cases, compost is used as a daily cover for landfills.

Yard Waste and Integrated Waste Management

Source Reduction: Short of cutting down trees and converting grass lawns to rock gardens, yard waste cannot be source-reduced in the same manner as most elements of the waste stream.

Source reduction for yard waste is generally considered to be backyard composting and "leave-it-on-the-lawn" (grasscycling) programs for grass clippings. Brush trimmings are most effectively shredded and used as mulch. Large-scale composting of leaf waste is a form of volume reduction resulting in loss of 40–75% of the original volume. In addition, approximately 50% of the weight is lost.

Composting (Recycling): Composting is generally defined as the controlled decomposition of organic matter by microorganisms into a humus-like product. Yard waste is organic and highly compostable. Nonmeat and nondairy food wastes can enhance the yard waste composting process. Composting is recycling for yard waste.

Incineration: Yard waste is combustible. It has a low Btu value (2,876 Btu per pound as compared to 4,500–5,000 Btus for a pound of MSW). Btu values for yard waste are heavily affected by the wetness of yard waste. Burning piles of yard waste is banned in a number of states due to potential air pollution and health problems (leaf smoke can create breathing problems for people who suffer from asthma, emphysema, chronic bronchitis, or allergies).

Landfilling: Yard wastes can decompose into methane. However, new EPA landfill regulations are designed to limit environmental degradation from methane production.

Yard Waste Composting Past and Present

Farmers have composted vegetable wastes with animal and human manures for thousands of years. Backyard composting of yard waste has long been practiced by many gardeners due to compost's attributes. Yard waste compost is not generally considered to be a fertilizer. However, it is a useful soil conditioner for improving texture, air circulation, and drainage. Compost can moderate soil temperature, enhance nutrient and water-holding capacity, decrease erosion, inhibit weed growth, and suppress some plant pathogens.

Curbside collection of yard waste for composting was not commonly practiced until the mid-1980s. As some states began to ban yard waste from landfill disposal, curbside collection of yard waste and source reduction programs encouraging backyard composting and grasscycling experienced a dramatic surge in growth. Currently, more than 2,000 communities have curbside collection of yard waste for composting programs. Nineteen states ban yard waste disposal. Another five states allow local bans or require composting programs if they are economically feasible.

The Yard Waste Composting Process

Compost, or "humus," is produced from the carbon content of yard waste while water and carbon dioxide dissipate into the atmosphere. Using a number of different techniques (windrows, static piles, in-vessel), the yard waste composting process generates energy and heat to destroy weeds and plant and human pathogens. To maintain aerobic conditions, most yard waste composting techniques require turning the compost to provide oxygen for the composting organisms. Temperature control of 132 to 140 degrees Fahrenheit is necessary for a sufficient length of time to kill off pathogens. Moisture content of 40–60% is required because yard waste that is too wet or too dry composts slowly. Finally, an adequate carbon to nitrogen (C/N) ratio is needed. Nitrogen is essential to composting. Grass is high in nitrogen, with a C/N ratio of 20:1. As a result, it must be balanced with leaf waste, which is low in nitrogen with a C/N ratio of 60–80:1. Food wastes are also high in nitrogen, while brush trimmings are very low.

The amount of time to produce compost varies from three to 18 months, depending on the process and amount of yard waste.

Limitations to Yard Waste Composting

Contaminants: Nonorganic materials (glass, metals, plastic bags, etc.) must be kept separate from leaf waste. In the past, pesticides could contaminate the compost, although many pesticides and herbicides marketed today are designed to break down rapidly in the soil. With the phase-out of lead gasoline for cars, lead contamination of compost is greatly reduced. Finally, tests around the country reveal little heavy metal contamination of yard waste.

Odor: Insufficient aeration or an improper C/N ratio are the main causes of intense odors.

Improper Operation: In addition to odor problems, failure to properly operate a compost pile can allow a fungus, asperillus fumigatus, to grow on composting leaves. This fungus can cause health problems.

Collection and Processing Costs: An EPA study of eight municipal yard waste composting programs reported collection costs as high as $80 per ton, with processing costs of up to $23 per ton. Only four of the eight programs studied showed revenue from the sale of compost, with one facility selling its compost for $25 per ton. Revenue did not exceed the cost in these composting programs. Generally, composting facilities must charge a tipping fee to cover costs. Avoided disposal costs are often cited as a key economic value in composting yard wastes. However, for avoided cost to be meaningful, the cost of collecting and processing yard waste must be less than the cost of landfill tipping fees.

Sources

Characterization of Municipal Solid Waste in the United States: 1992 Update, U.S. Environmental Protection Agency, Office of Solid Waste, 1992.

Composting Council, Alexandria, VA.

Yard Waste Composting (Environmental Fact Sheet), U.S. Environmental Protection Agency, Office of Solid Waste, 1991.

Measurement Standards and Reporting Guidelines, National Recycling Coalition, 1989.

Municipal Compost Management, Cornell Waste Management Institute, Cobb and Rosenfield, 1991.
The Solid Waste Management Problem, Council for Solid Waste Solutions, 1990.
Trash to Cash, Investor Responsibility Research Center.
Yard Waste Composting: A Study of Eight Programs, U.S. Environmental Protection Agency, Office of Solid Waste, April, 1989.

****

HOUSEHOLD HAZARDOUS WASTE

Household hazardous waste (HHW) consists of the residue (not the package) of many common consumer products. Components of these products include substances that could, if sufficient amounts are disposed of improperly, result in the release of potentially toxic substances. Due to the very low amount of hazardous substances in individual products, HHW is not regulated as a hazardous waste.

Examples of HHW include nail polish, toilet and drain cleaners, pesticides, small batteries, automotive products including used oil and batteries, and oil-based paints and paint thinner.

HHW Solid Waste Facts

Amount in Municipal Solid Waste (MSW): Unknown, but less than 1% of MSW by weight. In 1987, the U.S. EPA studied solid waste in two communities and determined that HHW constituted less than 0.35% to 0.40% of discarded MSW. A congressional report estimated that HHW discards could be as little as 300,000 tons, or 0.2% of MSW.

The average household in the EPA study discarded approximately 55 to 60 grams of HHW per week. This works out to approximately 4 to 5 pounds per person per year.

Volume in MSW: Unknown, but probably less than 1%.

Composition of HHW: In both locations in the EPA study, batteries outnumbered all other components of HHW, with "selected cosmetics" in second place. By weight, however, household maintenance products (paint, stain/varnish, etc.) were the largest component, followed by batteries in one community and automotive products (oil, antifreeze, etc.) in the other.

Recycling Rate: Most HHW products have a low to negligible recycling rate. The exception to this is lead-acid batteries, which had a 96.6% recycling rate in 1990 due to the value of the lead and the polypropylene battery casings.

HHW and Integrated Waste Management

Source Reduction: Source reduction (using substitute products without potentially hazardous components, or reducing the amount of these components in consumer products) offers the least expensive and most effective way to manage HHW.

Fourteen states have now passed legislation that essentially bans cadmium, lead, mercury, and hexavalent chromium from most consumer packaging. This legislation, known as the Toxics Reduction Law, is now functionally in effect throughout the U. S. for nationally distributed packaging.

Recycling: Many HHW products are recyclable, but programs to collect different types of HHW can be expensive, while getting only limited amounts of material. Lead-acid batteries, paint, used oil, antifreeze, and household batteries are the most commonly recycled HHW products.

Composting: Most HHW products are inherently uncompostable. HHW collection programs help reduce the amount of heavy metals in solid waste compost.

Incineration: The British thermal unit (Btu) value of HHW varies greatly, although most petroleum-based HHW will be relatively high in Btus. The low levels of HHW found in local waste streams and use of modern pollution control equipment should ensure that toxic elements are destroyed before they can be released to the atmosphere. Incineration is the most common method of handling used oil.

Land Disposal: Low levels of toxicity in HHW, the effect of wide dispersal of HHW in landfills throughout the U.S., and new environmental regulations should minimize any potential problems caused by landfilling HHW.

HHW Recycling Past and Present

Since 1980, 4,596 HHW collection programs were held. Most of these were one day programs. Every state has had at least one such program. In 1992, 857 programs collected HHW in 43 states.

HHW is most often collected on a special day, often called "Amnesty Day," during which local residents are instructed to bring specified types of HHW to a designated location. These programs may collect only one type of HHW (paint or antifreeze) or many different types. Some areas now use mobile collection vehicles to collect HHW from a variety of locations on a fixed schedule.

As interest in HHW recycling has increased, permanent drop-off sites or programs are becoming more common. In 1992, 128 permanent programs operated in 24 states. Permanent programs are defined as monthly collection at a fixed site or at a dedicated mobile facility. Permanent programs tend to collect greater quantities and more types of HHW than single-day collection efforts. One-third of the permanent programs are in California and Florida.

Finally, a small number of curbside recycling programs collect certain HHW items such as used oil and household batteries.

Limitations to HHW Recycling Variety in Types of HHW

The variety in the types of HHW products and their hazardous components create management and economic problems for HHW collection programs. Examples of products which pose individual handling and storage problems include: solvents (in products ranging from nail polish remover to paint thinner); metals (mercury in thermometers and small batteries, lead in lead-acid batteries, cadmium in many rechargeable batteries and some paint pigments, etc.); acids (found in toilet bowl cleaner and lead-acid batteries) and bases or alkalies (found in ammonia and some drain cleaners); and pesticides. Mixing acids and bases, for instance, can result in explosions.

Ironically, the most commonly collected item in HHW programs is nonhazardous, latex-based paint.

Some packages, such as motor oil cans and bottles, are excluded from MSW recycling programs because of HHW products in the package.

Cost

HHW collection programs are expensive to operate. According to a congressional report, participation in HHW programs is often low (less than 1% of the population), although individual contributions can be high (20–40 pounds per household because first-time participants often bring several years' worth of HHW). Unit costs can be as high as $18,000 per ton. Program costs can range from $2 per pound to as high as $9 per pound. In a more recent example, a very extensive program in the state of Washington costs $4 million per year for education and collection, yet only 5% of the population participates in the program.

Sources

American Plastics Council, Washington, D.C.

Characterization of Household Waste from Marin County, California, and New Orleans, Louisiana, U.S. Environmental Protection Agency, Office of Solid Waste, 1987.

Characterization of Municipal Solid Waste in the United States, 1992 Update, U.S. Environmental Protection Agency, 1992.

Facing America's Trash, Office of Technology Assessment, Washington, D.C., 1989.

Forty-five Questions and Answers about Composting, National Composting Program, Washington, D.C., 1993.

Hazardous Wastes from Homes, Enterprise for Education, Washington, D.C., 1991.

The National Listing of HHW Collection Programs, Waste Watchers Center, Andover, MA, 1992.

Waste Age's Recycling Times, January 12, 1993.

****

PLASTIC FILM BAGS

Plastic film is a thin-gauge plastic packaging medium used as a bag or a wrap. Examples include plastic grocery sacks, trash bags, drycleaner garment bags, carryout sacks, and plastic wrap including stretch wrap. Plastic film is less than 10 mils in thickness, with an average of 0.7 to 1.5 mils. A mil is 0.001 inch. Plastic film makes up 38% of plastic packaging.

Plastic film is a subset of a larger packaging group called flexible packaging. This larger category includes paper, plastic film, aluminum foil, and cellophane that is coated, printed, laminated, coextruded, and/or combined or used as a single material in packaging. Paper sacks and bags, microwaveable food packaging, and foil-laminated plastic pouches are examples of flexible packaging. Fifty-one percent of flexible packaging is plastic, 44% is paper, and 4% is aluminum foil.

Plastic Film Solid Waste Facts

Weight in MSW: U.S. EPA data show that 3.2 million tons of plastic film were generated in 1990 in the U.S. This is 1.7% of municipal solid waste (MSW) by weight before recycling.

The EPA tracks three types of plastic film in two waste product categories. These are trash bags (nondurable), bags and sacks (packaging), and wraps (also packaging). In 1990, 800,000 tons of trash bags, 900,000 tons of bags and sacks, and 1.5 million tons of wraps were generated.

After recycling, plastic film constituted 3.2 million tons or 2% of landfilled MSW.

The average American "generates" 25.5 pounds of these three types of plastic film in a year.

Volume: In 1990, the EPA estimated that plastic film occupied 9.5 million cubic yards of landfill space, or 2.4% of landfilled MSW by volume. The EPA estimates that plastic film products have a landfill density of 670 pounds per cubic yard.

A 30" × 42" × 48" bale of plastic film, baled on a horizontal baler, will weigh approximately 1,100 pounds.

Plastic film has a 1.2:1 volume-to-weight ratio.

Recycling Rate: In 1990, EPA data show "negligible" recycling rates for trash bags, a 3.2% recycling rate for bags and sacks (28,800 tons), and a 2% recycling rate for wraps (30,000 tons).

Recycled Content: Most plastic film has no post-consumer recycled content. However, pre-consumer recycled content is not unusual. California, the only state to require post-consumer content in plastic trash bags, now requires 10% post-consumer content in bags of 1.0 mil or greater, and in 1995, 30% in bags thicker than 0.75 mil. Most plastic trash bags are less than 1.0 mil in thickness. Recycled content can increase a bag's thickness by 50%.

Value: In March 1993, post-consumer plastic film pellets had list prices of 21–31 cents per pound. Clear, post-consumer low-density polyethylene (LDPE) film has the highest value.

Plastic Film and Integrated Waste Management

Source Reduction: Lighter in weight and less expensive than its competitors, plastic film is a prime example of successful source reduction. Producers and users of plastic film argue that its source reduction benefits greatly outweigh any recycling concerns.

Recycling: Because plastic film is thin and can be made from many different resins and colors, it is difficult to recycle. As a result, in 1992, the only state with a mandated plastic bag recycling rate changed the rate to a goal.

Composting: Plastic film does not compost. In an attempt to overcome this, degradable plastics are promoted for bags used in yard waste composting programs. Little evidence exists yet that degradable bags will work in this context.

Incineration: Plastic film is highly combustible, with a per-pound Btu value of 18,700 Btus for polyethylene, the most common film resin (compared with 4,500 Btus for a pound of MSW).

Landfilling: Plastic film does not degrade. As a result, degradable plastics have been proposed as a substitute. However, "degradable" bags will degrade slowly, if at all, in a modern landfill. Nonetheless, one state, Florida, requires all plastic bags to be degradable.

Plastic Film Recycling Past and Present

Clean, scrap plastic film generated by manufacturing and converting operations has a long history of recycling. These sources can supply sufficient amounts of clean raw material of a known resin type. In this case, the film is pelletized following a granulation or densifying process.

Post-consumer collection programs are more recent. They tend to focus on establishments that are large "generators" of plastic bags, such as grocery stores. In this case, the recycling process is more complex, requiring sorting, washing, and removal of contaminants as a first step. Curbside collection of plastic bags is rare.

Recycling Markets for Plastic Film

Markets include use of the pellet as a feedstock for plastic film, or for such diverse plastic products as tubing, sheet, bins, cans, LDPE bottles, and housewares.

Limitations to Plastic Film Recycling

Diversity of Plastic Resins: Most (80%) plastic bags and sacks and shrink-wrap are made from polyethylene, primarily LDPE. However, linear LDPE and high-molecular-weight HDPE also are used. The other 20% comes from polypropylene, polyvinyl chloride (used mostly for food wrap), and other resins. In addition, many films blend or coextrude two or more resins. Finally, individual product characteristics may create remanufacturing problems. Stretch wrap requires a "tackifier" so that the wrap can cling, yet this product quality is not desired in a bag.

Diversity of Colors: Probably half of plastic film is pigmented, while the other half is clear. In addition, bags often have print on them. As a result, the plastic must by separated by color and by printed versus nonprinted in order to be sold to the highest value end market.

Strict Raw Material Specifications: High value end markets for plastic film have strict raw material specifications, preferring clean, single-resin, clear pellets. Contaminants in plastic bag and film collection programs usually include food, paper receipts, labels, staples, non-plastics, and dirt.

Sources

Characterization of Municipal Solid Waste in the United States, 1992 Update, U.S. Environmental Protection Agency, Office of Solid Waste, Washington, D.C..

Measurement Standards and Reporting Guidelines, National Recycling Coalition, Washington, D.C., 1989.

Plastic Packaging Opportunities and Challenges, Testin & Vergano, SPI, Washington, D.C., 1992.

Modern Plastics, January, 1993.

Plastics News, March 22, 1993.

Plastics Recycling Issues, No. 2, May 1992, Stan Norwalk, Norwich, CT.
The Role of Flexible Packaging in Solid Waste Management, Flexible Packaging Association, Washington, D.C., 1990.
Tom Tomaszek, North American Plastic Recycling, Ft. Edward, NY, 1993.

****

SCRAP TIRES

More than 264 million tires were sold in the U.S. in 1990. In the same year, more than 242 million scrap tires were generated. Almost 80% of scrap tires come from passenger cars, with the remainder coming primarily from trucks. Over 50% of the rubber consumed in the U.S. is used to make rubber tires. The weight of tires varies: on average, one passenger car tire weighs 20 pounds; one light truck tire weighs 35 pounds; and one semi-truck tire weighs 105 pounds. The U.S. EPA estimates that between 2 and 3 billion scrap tires are stockpiled in the U.S.

Weight in MSW: In 1993, according to the EPA, 3.4 million tons of scrap tires were generated in the U.S. This is approximately 1.6% of municipal solid waste (MSW) by weight.

In 1993, 2.9 million tons of scrap tires were landfilled, stockpiled, or illegally dumped. According to the EPA, scrap tires are 1.8% of landfilled MSW by weight.

Volume in MSW: EPA landfill volume data do not include tires.

Recycling Rate: In 1990, 26 million scrap tires (10.7%) were burned for their energy value, 16 million (6.7%) were recycled, and 12 million (5%) were exported. An additional 33.5 million scrap tires were retreaded in 1990 and 10 million were reused.

Value: Negative. In most cases, generators pay a tip fee to scrap tire markets. The tip fee varies regionally and by tire size.

Recycled Content: New tires contain no more than 2% recycled rubber. Retreads contain 75% recycled content.

Scrap Tires and Integrated Waste Management

Source Reduction: Designing tires for longer life, reuse of used (but still usable) tires, and retreading are the primary source reduction options for scrap tires. Over the past 40 years, improved manufacturing techniques have doubled the useful life of tires. Forty-thousand-mile tires are now commonplace.

Retreading is applying a new tread to a worn tire with a good casing. Retreading is on the decline because of the low price of new passenger car tires and misconceptions about the quality of retreads. However, the number of truck tire retreads is increasing, due to the high cost of truck tires. Over 1,700 retreaders operated in the U.S. and Canada in 1991.

Recycling: Scrap tires can be recycled as whole or split tires or as crumb (ground) rubber. Whole tire uses include artificial reefs and playground equipment. Split tire uses include floor mats, belts, and dock bumpers. Crumb rubber uses include rubber and plastic products such as mud guards, carpet padding, tracks and athletic surfaces, and rubberized asphalt. Crumb rubber is by far the biggest recycling market for scrap tires.

Composting: Scrap tires do not compost. However, 2-inch square shredded tire chips can be used as a bulking agent in composting wastewater treatment sludge. The tire chips must be removed from the compost before it is sold, but they can be reused in the composting process.

Incineration: Scrap tires are highly combustible with a 12,000 to 16,000 British thermal unit fuel value per pound, which is slightly higher than coal. Whole tires can be burned or the tires can be shredded into tire-derived fuel (TDF). Hog fuel boilers for pulp and paper mills are the biggest users of scrap tires, followed by cement kilns and combustion facilities designed to burn TDF.

Landfilling: Unlandfilled scrap tires can create a public health problem as a mosquito breeding area. Landfilling single tires can pose problems if the tires fail to compress within the landfill and resurface. As a result, six states ban the disposal of tires, while another 21 ban the disposal of whole tires. Disposal bans passed without recycling or incineration options can lead to stockpiled whole tires and tire pile fires.

Scrap Tire Recycling Past and Present

Historically, retreading has been the main form of scrap tire recycling. However, with the decline of the retreading industry, and the high cost of disposing of scrap tires, increasing numbers of tires have been illegally dumped. A recent EPA report identified TDF and rubberized asphalt as the best solutions for scrap tire recycling.

Limitations or Barriers to Scrap Tire Recycling

TDF and Combustion in Cement Kilns: Tire combustion faces the same NIMBY (Not In My Back Yard) opposition that plagues all combustion technologies. Additionally, in many cases, use of TDF or whole tires as a fuel requires an additional, lengthy permitting process for existing boilers. Competing fuels may also be less expensive. Finally, individual boilers have different fuel specifications. Some can burn whole tires, others need finely shredded, dewired tires.

Rubberized Asphalt: Rubberized asphalt comes in two forms. One is using tire crumbs or chips as 3% of the aggregate in asphalt (also known as Rubber Modified Asphalt Concrete, or RUMAC). The other is blending crumb rubber as 15–25% of the asphalt binder. The binder is used as a sealant.

Using crumb rubber as an additive in pavement is a proven technology with tremendous end market potential. Yearly U.S. asphalt needs are 10 times the annual supply of scrap tires. The 1991 Intermodal Surface Transportation Efficiency Act requires that 5% of federally-financed asphalt laid in a state must contain recycled rubber. In 1997, the minimum percentage is 20%.

Rubberized asphalt is not without drawbacks, however. The initial investment cost is high, often twice normal highway repair cost, although rubberized asphalt can double pavement lifetime. In addition, rubberized asphalt does not have long-term performance test results, may not itself be recyclable, lacks national specifications, and is a change in the recipe for a product with long-established specifications.

Value: Most products made from scrap tires are low value products with, in many cases, higher costs than competing products.

Sources

Characterization of Municipal Solid Waste in the United States: 1992 Update, U.S. Environmental Protection Agency, Office of Solid Waste, 1992.
Facing America's Trash, Office of Technology Assessment, Washington, D.C., 1989.
Markets for Scrap Tires, U.S. EPA, October, 1991.
Measurement Standards and Reporting Guidelines, National Recycling Coalition, Washington, D.C., 1989.
Getz, N., and Teachey, M.F., Options in scrap tire management, *Waste Age*, October, 1992, p. 81.
Scrap Tire Management Council.
Tire Retread Information Bureau.
Robinson, C.L., Tire shredding, *Waste Age*, October, 1991, p. 75.

****

ASEPTIC BOXES/MILK CARTONS

NOTE: This Profile covers aseptic boxes and paper milk cartons because they are collected together in recycling programs, but statistics are given separately for each product.

Aseptic boxes, more commonly known as drink boxes, are used for fruit juices and milk. Aseptic processing involves a high-temperature/short-time treatment in which liquid products are heated quickly to a temperature at which sterilization occurs. The product is then cooled and placed in a sterile container.

By weight, aseptic boxes are 70% paper (used for stiffness and strength), 24% polyethylene (used in four different layers to seal the package liquid-tight), and 6% aluminum foil (used as a barrier against air and light). Aseptic boxes can protect beverages for a year or more without the need for refrigeration.

By weight, milk cartons are 80% paper and 20% polyethylene. Milk cartons, also known as gable top cartons, are the most common form of polycoated paper packaging. Fruit juices (not covered in this Profile) are also found in gable top boxes. Also not covered in this Profile are other forms of polycoated paper packaging such as frozen food boxes, round ice cream cartons, and microwaveable dinner cartons.

Weight in MSW: According to the U.S. EPA, 500,000 tons of milk cartons were generated in the U.S. in 1990. This is 0.3% of municipal solid waste (MSW) by weight. Approximately 35,000 to 40,000 tons of aseptic boxes were generated in 1992, which would be less than 0.03% of MSW by weight.

The average milk carton weighs 1 oz. The average aseptic box weighs 10 grams (approximately 90,000 drink boxes in a ton). The average American "generates" approximately one-third of a pound of aseptic boxes each year and 4 pounds of milk cartons. American consumers use 3 billion aseptic boxes and 17 billion milk cartons each year.

Recycling Rate: The U.S. EPA estimated a "negligible" 1990 recycling rate for milk cartons. The U.S. EPA did not estimate a drink box recycling rate.

Amount Not Recycled: In 1990, 500,000 tons of milk cartons and 25,000 tons of aseptic boxes were disposed of in landfills or incinerators. This is 0.33% of discarded (nonrecycled) MSW.

Volume in MSW: In 1990, the U.S. EPA estimated that milk cartons occupied 1.2 million cubic yards of landfill space, or 0.3% of landfilled MSW by volume. Collected aseptic boxes and milk cartons have a density of 80–100 pounds per cubic yard. Landfilled milk cartons have a density of 820 pounds per cubic yard.

Value: The long paper fibers found in both containers are quite valuable, with some markets paying in excess of $100 per ton for aseptic boxes and milk cartons.

Markets: End-markets include printing and writing paper, tissue, and paper towels.

Aseptic Boxes, Milk Cartons and Integrated Waste Management

Source Reduction: Aseptic boxes are a classic example of source reduction of packaging. Lightweight (with consequent savings in transportation costs) and unbreakable, drink boxes also save energy because they do not require refrigeration to preserve the product for a year or longer.

Recycling: Aseptic boxes and milk cartons are not commonly recycled. As a result, recycling advocates succeeded in banning aseptic boxes for most products in one state—Maine—for several years; the ban was finally lifted in 1994. While source reduction is at the top of the solid waste management hierarchy in most states, aseptic box recycling receives more attention than its source reduction benefits.

Composting: The mix of materials in both packages creates composting problems. Aluminum foil and polyethylene are noncompostable. Composting operations will attempt to remove both by mechanical means.

Incineration: Plastic and paper have relatively high Btu values (19,000 Btu per pound for polyethylene and over 7,000 Btu per pound for paper, as compared to 4,500–5,000 Btu for a pound of MSW). The aluminum foil in the aseptic box, however, will not combust and will end up as a residue in the ash.

Landfilling: The paper in both packages will degrade very slowly in a modern landfill. The plastic and the aluminum foil will not degrade.

Aseptic Box and Milk Carton Recycling Past and Present

Pre-consumer (converter scrap) polycoated paper fiber that does not require deinking has always been recycled because of the high value of the paper fiber. Its strength and brightness make it ideal for high-value products. However, larger or heavier containers first received the attention of post-consumer recycling programs. Due to the Maine ban and consumer pressure, the aseptic packaging industry has piloted recycling programs in schools and at the curbside. The industry estimates that almost 1,500 schools and over 1 million households participate in recycling programs for these packages.

The Aseptic Box/Milk Carton Recycling Process

After collection, these packages are hydropulped to separate the paper fibers from the plastic and aluminum foil. Because graphics are printed on the outer polyethylene layer, the paper fiber does not require deinking. The plastic can be burned for steam at a paper mill. If local markets exist, the plastic/foil mix can be used for plastic lumber.

Limitations to Aseptic Box/Milk Carton Recycling

Weight: Aseptic packaging's source reduction strengths are its recycling weaknesses. Because aseptic boxes are so light, the aseptic industry promotes collection of milk cartons along with aseptic boxes. Milk cartons are far more plentiful than aseptic boxes and have the same high value paper fibers.

Contaminants: Failure to empty milk cartons will result in milk sugars remaining in the box. In a hot or humid climate, these sugars will attract bugs. Liquids and the polyethylene coating can also create baling problems as the packages squish together.

Collection and Processing Costs: Due to the limited number of programs, good cost data do not exist. Aseptic boxes and milk cartons are lightweight (which should lead to higher costs), but they also have a denser volume than many recyclables (which should lead to lower costs).

Sources

Aseptic Packaging Council, Washington, D.C.

Characterization of Municipal Solid Waste in the United States: 1992 Update, U.S. Environmental Protection Agency, Office of Solid Waste, 1992.

Facing America's Trash, Office of Technology Assessment, U.S. Congress, 1989.

Maximum Benefit, Minimum Waste, Aseptic Packaging Council, Washington, D.C.

Paperboard Packaging Council, Washington, D.C.

Hood, P., Polycoated paper recycling: the expansion from pre- to post-consumer, *Waste Age*, April, 1993.

The Solid Waste Management Problem, Council for Solid Waste Solutions, Washington, D.C., 1990.

Walden's Fiber and Board Report, October 14, 1991.

Waste Wise, Aseptic Packaging Council, Washington, D.C., 1991.

****

MAGAZINES AND CATALOGS

Most magazines and catalogs are printed on coated, groundwood paper. Clay, by far the most common coating, is used to help smooth the paper surface and to create an optimum surface to which glossy inks can adhere. Groundwood is the same kind of paper used for newspapers.

A two-sided coated paper sheet used for magazines will normally have 30–35% clay and filler, and 65–70% paper fiber.

Approximately 16 billion magazines were distributed in the United States in 1989. Of these, 66% go to subscribers, 17% are sold in newsstands, and 15% are returned as unsold.

Weight in MSW: In 1990, according to the U.S. EPA, 2.8 million tons of magazines were generated as municipal solid waste (MSW) in the United States. This is 1.4% of MSW by weight. (U.S. EPA magazine data does not include catalogs.)

The magazine industry estimates that 4 million tons of magazines were produced in 1990. Another 1.9 million tons of catalogs were produced in 1990.

On average each American "generates" 47 pounds of magazines and catalogs each year.

The "average" magazine weighs one-half pound.

Recycling Rate: The U.S. EPA estimated a 10.7% magazine recycling rate for 1990. This is 300,000 tons of magazines. Due to increased demand, 1,000,000 tons were recycled in 1993.

Amount Not Recycled: In 1990, according to the U.S. EPA, 2,500,000 tons of magazines were disposed of in landfills or waste-to-energy facilities. This is 1.5% of discarded (nonrecycled) MSW.

Volume in MSW: In 1990, the U.S. EPA estimated that magazines occupied 6.3 million cubic yards of landfill space, or 1.5% of landfilled MSW by volume. Landfilled magazines have a density of 800 pounds per cubic yard.

Value: Most processors charge a tipping fee or take magazines for no cost. Most end-markets pay more for magazines than for newspapers.

Markets: Newspaper deinking mills using flotation technologies to remove ink from newspaper fiber are the primary market for magazines. Clay particles help adsorb ink and then rise to the surface with air bubbles in the flotation cell. The most common deinking mixture in flotation systems is 30% magazines to 70% newspaper. As much as 2 million tons of magazines will be needed to help deink newspapers by 1995.

Recycled content in coated paper is a recent phenomenon. A 1993 survey showed that 21% of the responding magazine companies use some recycled-content paper. The survey also showed that North American use of recycled coated groundwood paper would rise by 118% in 1994.

Other markets include traditional mixed wastepaper markets such as container board and tissue paper.

Magazines and Integrated Waste Management

Source Reduction: Lighter-weight and smaller-sized paper can be used. Total magazine production decreases primarily when an economic recession causes sales to decline or causes fewer pages to be printed due to fewer ads and fewer stories. Trends in consumer purchasing habits are leading to increased production and distribution of catalogs.

Recycling: In the past, coated paper was hard to recycle because deinking technologies recovered more clay than paper fiber. New flotation deinking technology has turned a disadvantage into an advantage.

Composting: Magazines can be composted if shredded properly; however, the clay coating will resist composting.

Incineration: Magazines are combustible. However, their clay content gives them about the same Btus per pound as MSW (4500 to 5000 Btus for a pound of MSW). Burning coated paper will create more ash than burning other forms of paper.

Landfilling: Magazines, like most materials, will degrade very slowly in a modern landfill.

Magazine Recycling Past and Present

Traditionally, magazine recycling has been limited to relatively clean and easy to collect sources such as printer's pressroom waste and unsold copies. As newspaper deinking mills increasingly need magazines for their clay content, drop-off and curbside programs will increase.

However, curbside programs are relatively new and usually collect small amounts of magazines. These low rates are caused by the tendency of people to keep magazines longer than newspaper, and the failure to adequately publicize the addition of magazines to a recycling program. As magazine recycling programs become more common, the amount recycled should increase.

Magazine Grades for Recycling

The Paperstock Institute specifications for magazines call for "dry, baled, coated magazines, catalogs and similar printed material. May contain a small percentage of uncoated news-type papers." Prohibited materials are limited to 1% and out-throws to 3% (PSI Grade # 10).

Limitations to Magazine Recycling

Contaminants: Some magazines are contaminated for recycling by materials used to produce or distribute the magazine. These contaminants include ultra-violet cured inks, pressure-sensitive adhesives, water soluble glue bindings, plastic bags, and metallic or plastic inserts.

Processing: Unshredded magazines can be slippery and hard to compress. When baled, stacks tend to slide and fall apart. As a result, more care and time must be taken in baling whole magazines. Shredded magazines do not offer as many problems.

Collection and Processing Costs: Most deinking mills require that newspaper and magazines be kept separate. Adding magazines to a collection program may require an extra bin on the collection truck or additional processing capacity. Either option can lead to increased costs.

Sources

Characterization of Municipal Solid Waste in the United States: 1992 Update, U.S. Environmental Protection Agency, Office of Solid Waste, 1992.
John Davis, Davis Recycling Consultants, Colleyville, TX.
Institute of Scrap Recycling Industries, Washington, D.C., 1993.
Janet Malloch, Smurfit Newsprint Corporation, Oregon City, OR, 1993.
Bill Moore, Thompson-Avant International, Atlanta, GA, 1993.

Paper Recycler, December, 1993.
Recycling in America, Debi Kimball, 1992.
Recycling Times, December 14, 1993.
Nancy Risser, Risser and Associates, New York, N.Y.
The Magazine Wastepaper Stream: Addressing the Challenge, Magazine Publishers of America, Risser and Associates, 1990.
Waste Age, The Emerging World of Deinking, June, 1992.

****

HOUSEHOLD BATTERIES

NOTE: This profile does not cover wet cell lead-acid batteries such as car batteries.

A battery is a device in which the energy of a chemical reaction can be converted into electricity. Household batteries include primary batteries, which cannot be recharged, and secondary (rechargeable) batteries. Household batteries are available in many sizes including button, AAA, AA, C, D, N, and 9-volt.

Battery types (and examples of household products using them) include alkaline and zinc-carbon (flashlights, radios, toys, calculators); nickel-cadmium (portable rechargeable products); silver oxide (calculators, watches); zinc-air (hearing aids); mercuric oxide (hearing aids); small-sealed lead-acid (some rechargeable products); and lithium (computers and cameras).

Alkaline batteries had 63% of consumer product sales in 1992, followed by zinc carbon with 19%, and nickel-cadmium with 13%. Silver oxide and zinc-air each had about 2% of the market, with the remaining 1% split among the other battery types.

Weight in MSW: In 1991, 3.8 billion household batteries, or 145,000 tons, were incinerated or landfilled in the United States. This is less than 0.1% of discarded municipal solid waste by weight.

The average American discards 15.8 household batteries per year. This is 1.16 pounds or 530 grams.

The average weight of different types and sizes of household batteries varies dramatically. Silver oxide button batteries weigh 0.96 grams on average; size C alkaline batteries average 57.7 grams, while

AA alkalines average 22.9 grams; and nickel-cadmium batteries average 44.0 grams.

Recycling Rate: The U.S. EPA did not estimate a 1990 recycling rate for household batteries. Due to a limited number of programs collecting batteries, the recycling rate is very small.

Volume in MSW: The U.S. EPA did not estimate the density or volume of household batteries in landfills in 1990. Assuming that batteries have equal volume and weight percentages of MSW, household batteries occupied approximately 370,000 cubic yards of landfill space (0.09%) in 1990.

Hazardous Constituents: An EPA study estimated that in 1989, 52% of the cadmium and 88% of the mercury in MSW came from household batteries. Nickel-cadmium and mercuric oxide batteries fail the Toxicity Characterization Leaching Procedure for cadmium and mercury, respectively. This means household batteries from nonhousehold sources would be considered hazardous waste upon disposal. In addition, California regulates nickel, silver, and zinc as toxic metals.

Value: While some batteries such as silver oxide button batteries can be sold for their metal content, most end markets charge a fee to take batteries. Alkaline and zinc-carbon, the most common batteries, have the least valuable metals and the fewest end-markets.

Markets: At least two American companies use household batteries as a source of silver, mercury, nickel, cadmium, and iron.

Household Batteries and Integrated Waste Management

Source Reduction: Rechargeable batteries are a source reduction option because their rechargeability avoids the need to purchase primary batteries. However, rechargeable batteries do not last forever. Failure to recycle them will increase the amount of cadmium in the waste stream.

Battery manufacturers successfully reduced the amount of mercury in dry cell batteries by 75% from 1984 to 1989. Alkaline battery manufacturers will not intentionally use mercury, and mercuric oxide batteries will be phased out by the mid-90's.

Recycling: Household batteries are not recycled back into batteries. Instead, they are collected for metals reclamation. The low value of most battery metals and the high cost of collecting and processing, lead to a low battery recovery rate.

Composting: Due to the heavy metals in batteries, household batteries are not compostable.

Incineration: Batteries are combustible. However, concern over mercury and other heavy metals in incinerator emissions and ash has caused several states to ban incineration of household batteries.

Landfilling: New EPA landfill regulations are designed to limit any environmental problems caused by the eventual degradation of heavy metals found in batteries generated by households.

Household Battery Recycling Past and Present

Curbside recycling of batteries is primarily found in the few states that have either mandated collection or banned disposal of some types of household batteries. Batteries can also be collected in drop-off programs, although these programs tend to have lower participation rates than curbside collection.

After collection, batteries with valuable metals go to recyclers who use thermal or chemical processes to separate out the metals. Batteries without those metals are usually sent to a hazardous waste disposal facility because they were probably mixed with batteries from nonhousehold sources.

Limitations to Household Battery Recycling

Hazardous Waste Regulations: A program that mixes batteries from household and nonhousehold sources is liable for the full array of federal and state hazardous waste collection, handling, and disposal requirements. Some states, however, exempt batteries from small businesses from these requirements.

The EPA has proposed regulations to encourage recycling by easing the regulatory requirements for battery recyclers.

Cost: Limited data exist on collection and handling costs. However, costs will be high due to the small size of most household batteries and limited amounts collected by operating programs. One study of four programs showed cost ranges of $1,840 to $4,840 per ton. The programs collected from 7% to 18% of the available batteries.

Battery Accessibility: Many products using rechargeable batteries were designed with the battery encased in the product. As a result of legislation passed in several states, products using rechargeable batteries are now designed to ease removal of the batteries. This will improve their recyclability.

Sources

Characterization of Municipal Solid Waste in the United States, 1992 Update, U.S. Environmental Protection Agency, Office of Solid Waste, 1992.
David Hurd, R2B2, New York, N.Y.
Getting a Charge Out of the Wastestream, Council of State Governments, 1992.
Household Battery Waste Management Study, California Integrated Waste Management Board, 1992.
Outlook for Recycling Large and Small Batteries in the Future, EG&G, Idaho, 1986.
Portable Rechargeable Battery Association, Atlanta, GA.
Report on Dry Cell Batteries in New York State, NYS Department of Environmental Conservation, 1992.
Used Dry Cell Batteries: Is a Collection Program Right for Your Community?, U.S. EPA, 1992.
Wisconsin Household Batteries Waste Management Study, Gershman, Brickman, Bratton, Falls Church, VA, 1992.

****

CODA: THE WHYS AND HOWS TO QUALITY RECYCLABLES

By *Michael Misner*, former market analyst for *Recycling Times*, Washington, D.C.

Mike Anderson of Garbage Reincarnation, Inc., in California was losing money. Loads of his processed glass were being rejected after 15 years in the business without as much as a denial, and suddenly the glass manufacturer rejecting his material said it was dirty and worthless.

And then there's Randy Ward of the Chesapeake Paper Mill in West Point, VA, who more than once has opened a bale of old corrugated cardboard to find spent generators and other junked machinery. He has since stepped up examination of all incoming material to track such contaminants and close those accounts.

In the past, clean recyclables fetched one price while dirty recyclables, or what we might consider less-than-quality material, earned another price. Although that price was lower, the recyclables were never completely rejected. It was the nature of the business, but it has changed (see Table 1.1).

Ensuring quality has become an integral part of a recycling operation's competitive formula as end-users become stricter. End-users know poor quality feedstock costs, and that it carries the potential to cheapen their products.

Less-than-quality material may still be accepted, but in the long run it will pay recyclers to produce a quality product—not only to get the best prices, but also to secure a healthy market environment.

Paper

Old newspapers (ONP) are one of the most commonly collected recyclables from consumers, and also the commodity where consistent quality is most elusive. Paper mills cringe at the thought of glue or gummed material such as labels and adhesive tapes mucking up their expensive machinery. ONP is the largest culprit harboring these items, according to Garden State Paper's Frank Lorey (Saddlebrook, NJ).

Such stuff forms balls of goo called "stickies," which stick sheets of new paper together and create a continuous tear on the presses that can ruin an entire roll of paper. "Keeping stickies out is our No. 1 quality concern. They plug up the head box, cause weak sheets of paper . . . a mess all the way down the line," said Charlie Poland of Environmental Control (Arlington, VA).

"Sophisticated screening equipment has been developed for effective removal of contaminants—including stickies—but no approach is perfect, and removal efficiency is about 90%," Lorey said. In the large quantities many mills deal with, even 10% contaminants means tons of wasted material.

Quality ONP must also be dry, relatively young, and nonsunburned. Sunlight damages ONP brightness and strength qualities, while water rapidly decays stored paper. Also, ONP over six months in storage gains brittleness and loses fiber strength. Paper inventories should be readily rotated to avoid these contamination problems.

Table 1.1 Stuff Nobody Wants In Their Recyclables

Should not be in collected post-consumer old NEWSPAPERS:
Gummed material such as stamps, stickers, labels,
Envelopes with plastic windows, plastic bags, other plastics
Glossy material like magazines or junk mail
Metal objects such as staples, screws, coins, cans, etc.
Rubber, such as rubber bands, erasers
Food wrappings, paper towels

Storage points of ONP:
Keep dry
Out of direct sunlight
Rotate old material out as soon as possible
No twine or string
No dirt or debris

Should not be in collected post-consumer OFFICE PAPERS:
Post-It notes
Colored paper
Carbon paper
Envelopes
Labels, stickers, photos, glossy material
Sometimes laser paper
Food wrappings, paper towels
(Source: Weyerhauser's We-cycle Office Paper program)

Should not be in collected post-consumer or post-industrial OCC:
Waxed board
Food waste
Styrofoam packaging peanuts
Plastic liners
Adhesives

Should not be in collected post-consumer GLASS:
Ceramic cups, plates, wine caps, flower pots
Crystal
Light bulbs
Mirrors and window glass
Heat resistant ovenware
Drinking glasses
Broken glass
Dirt
(Source: Mid-Atlantic Glass Recycling Program)

Should not be collected in post-consumer PLASTICS:
Other plastics, unless a mixed load
Mixed colors, unless buyer wants them
Metal, dirt, glass

Hints to identifying PVC:
PVC has a number 8 on the bottom
Crystal clear or colored, but shiny surface

Should not be collected in post-consumer UBCs:
Moisture
Lead
Other metals, paper, glass, and plastic
Watch for very specific specs on bale size, methods of packaging, etc.; depends on the smelter

Strength or runability is important to mills because a paper tear stops the presses. Any shutdown to mend a tear can cost a mill thousands of dollars per minute as highly specialized equipment sits idle. Lorey said clean ONP is actually higher in tear resistance than virgin newsprint, but not if the paper is improperly handled and stored.

Coated papers such as magazines or inserts are contaminants to newsprint recycling mills using a washing system, while mills using a floatation system actually want about 30% of their mix to be these glossy papers. In the floatation system, the clays from the coated papers help absorb and remove ink particles. Knowing whether or not your mill wants any coated paper is part of the consistent communication necessary between processors and individual mills when dealing with recyclables.

And dealing with ONP is here to stay. Twenty-four percent more quality ONP will be required in 1995 than in 1990 as production of recycled newsprint is expected to increase, according to a 1990 leadership opinion survey of newsprint producers by the Wirthlin Group headquartered in McLean, VA.

According to the 26 newsprint producers interviewed in the Wirthlin survey, a new deinking mill costs up to $80 million. Before investing in such capital, mills need assurance of large, consistent quantities of quality ONP. Seventy percent of Canadian newsprint producers in the report said their primary concern about recycling was "adequacy and stability of supply." Thirty-one percent of producers ranked supply quality as one of their major concerns.

The worries may be justified. The leadership report also said, "collection systems are not yet producing quality raw materials, estimating that between 20 to 40% of ONP is not usable because it contains contaminants such as plastics, rubber bands, chewing gum, etc."

Although most mills are just starting to put their toes in the water, some mills are coming on-line with some serious recycling capacity, despite concerns. Together with statewide recycling programs and busy processors, mills may be giving ONP new marketable life.

Old corrugated cardboard (OCC) and computer printout (CPO) are also marketable, but have their own quality problems. One source in Ohio said his primary problem with OCC is foreign cardboard of lesser quality. Mainly from oriental sources, the material is a "bogus recycled sheet that mills don't want as feedstock," he said.

Most of this poor quality OCC has been recycled once before. Department store packaging material generates the highest percentage of this unwanted form of OCC. Supermarkets' OCC, the other major OCC generator, has other quality considerations.

"My main problem is OCC from supermarkets: Too much food waste. We've had to reject many loads because of it," said Tim Epps of Quick Service Metals (Tulsa, OK). Waxed boxes used to hold damp produce are another source of OCC contamination, according to processors.

Max Marsh of the Rock Tenn Corporation (Lynchburg, VA), which processes only OCC and mixed waste, said 5% of his production is spent dealing with plastic contaminants such as Styrofoam packaging peanuts. Although screening equipment catches most of the material, cleaning consumes valuable production time.

On the other end, Chuck Stone from Recycled Fibers of Ohio (Moraine, OH) sees mills increasing their quality standards. They are setting up a statistical process of OCC quality control with periodic testing of characteristics such as fiber strength, he said. Most mills currently [in 1991] will not accept over 5% contamination of this stuff.

High grade office paper has its own special contaminants. Laser CPO and other toner image papers are considered the main contaminants to office paper recycling programs.

Mixed office paper grade prices have suffered from increased laser contamination. Because the toner actually fuses to the fiber with heat, laser-rich paper is very difficult to process, and creates brown spots when it is pulped. At a recycled tissue mill, for example, this shows as blemishes on new tissue.

Determining between laser-free and laser CPO is a growing problem for dealers of high grade office paper. Stone said microscopic examination is the best way to differentiate, but that kind of scrutiny is impossible, and it is unreasonable to expect a $5/hour handpicker to tell the difference. Other serious contaminants are pressure-sensitive labels, Post-It notes, and return labels.

Many mills take laser-free CPO only if it is a big, well-known account. Stone warned that "communication on quality between customer and supplier is critical."

The most commonly recommended remedy for contaminants in ONP, OCC, and CPO is persistent education. Going back to accounts and educating the public is the best way to deal with the problem, recyclers say.

Glass

Glass manufacturers also bemoan quality. Their concerns are ceramics, dirt, and breakage. Consumers need to be told that a ceramic jar which looks and feels like glass is unwanted in a glass load.

In the glass oven, ceramic does not melt at the same temperature as glass. That means tiny bits of ceramic are lodged in the new glass and serve as points of imperfection. Bubbles form that increase the new container's brittleness, thereby decreasing its quality. Sometimes such impurities cause explosions inside the furnace. Those explosions not only damage equipment but endanger lives.

In 1990, the issue of quality in glass affected usually stable glass prices. In the Midwest and West, prices dropped $10–20 per ton for clear, green, and brown glass, with quality targeted as the main reason.

Current specifications are taken more seriously, especially as glass manufacturers are using more recycled material where, again, the larger the percentage of recycled material used, the greater the amount of contamination for even a small, allowable percentage.

Austin Fiore of Owens-Illinois in Pennsylvania said at the New York State Legislative Commission on Solid Waste [in 1991] that glass specifications are growing tighter because "the smallest amount of debris, especially ceramic" can foul new glass.

"It is because of supply limitations on clean cullet that the industry average for recycled content in a glass bottle remains at 30%," Fiore said. In areas where manufacturers get cleaner material, the individual average is as high as 60%.

Yet ceramics and such debris may not be a problem in the near future, at least in Canada. Consumers Glass in Toronto completed experiments on 350 tons of post-consumer glass contaminated with ceramics.

By grinding the whole batch to a fine sand-like material using powerful mineral crushing equipment, the ceramic particles were made harmless. Consumers Glass is seeking bidders to build a new plant based on this technology, sources say. Although not encouraging ceramic content, the process increases glass' recyclability, despite ceramic contamination.

Breakage in general is discouraged by glass makers because it is difficult to monitor contamination in thousands of glass fragments. Tom McCaughey of McCaughey Standard (Pawtucket, RI), was having problems marketing his glass to New England CRInc. because of breakage. The less-than-quality glass he collected for recycling was landfilled. McCaughey solved his problems by switching trucks. He found trucks with separate bins decreased breakage and increased quality. But many manufacturers realize densification is necessary to make long shipments affordable and consider color-sorted, broken glass acceptable.

Consumers Glass in Toronto has also found that curbside color separation into brown, green, and clear glass decreases contamination. Drivers separating at curb make sure no ceramics go into any of the three glass receptacles.

Aluminum Cans

The most successful recyclable is aluminum used beverage cans (UBC). Most quality problems such as dirt, plastics, and steel cans can be separated out with relative ease.

Magnets pull out steel cans as aluminum cans move down a conveyor. Dirt and plastics can be shaken out or left behind with various uses of gravity.

Processors and end-users cannot stop talking about moisture as a quality problem, however. Moisture content in cans has been a hot issue since major buyers of aluminum cans began discounting loads for too much of it.

"Naturally, we don't want to pay for water," said Reynolds Aluminum's spokesperson Lou Anne Nabhan. Other reasons for the crackdown on moisture are safety and economics. Nabhan said moisture can cause dangerous explosions in the furnace, and that moisture overall decreases recovery rate.

Most manufacturers like Reynolds have set up their own standards for moisture content. "Each mill has to make a decision based on what they think is fair," said Linda Poetter of Golden Aluminum (Golden, CO).

Samples are taken from the front, middle, and end of loads. They are weighed, baked, and weighed again. The difference between the first and second weight determines how much moisture was in the sample and that percentage may be deducted from the original weight of the load.

Reynolds deducts for anything over 4% moisture content, but will only deduct back to an industry standard of 2% allowable content. For example, if a sample from a load is 5% moisture by weight, a 3% deduction from the original weight will be levied and the processors will be paid 3% less by weight than they assumed they would earn.

"Moisture is a reality. Cans were designed to hold it, and they do it well," said Tom Mele of Connecticut Metal (Monroe, CT). He said for 15 years moisture was not a problem, and now it is. Mele said 80–90% of processors can continuously provide less than 6% moisture content, but 4% is much tougher. Although the debate continues, moisture content as a quality issue is an everyday reality for processors.

"To avoid problems, do your job of inspection, run your cans across a magnet, shred, or remelt the material," suggests one source from a secondary aluminum smelter.

To avoid moisture, cans should not be stored outside, especially in the winter. Densifying the UBCs is not recommended by industry sources because it locks moisture into the bale and after that, regular methods of moisture reduction are useless.

Cans should be heated if possible, to dry excess liquid. Shredded cans are said to have the least amount of moisture because the process disperses moisture well and exposes more surfaces to air drying. However, many operations cannot afford to own and operate a shredder. In such cases, a wire tumbler is recommended to shake extra liquid off the cans.

Plastic

The main quality problem for plastics is other plastics. To a plastic manufacturer, mixing resins is like putting broccoli in an apple pie and expecting it to still taste and look like an apple pie.

"PVC (polyvinyl chloride) is, without a doubt, the No. 1 problem," said Bob Dastou of Wellman Inc. (Shrewsbury, NJ). Wellman is the largest PET (polyethylene terephthalate) recycler in the country, handling millions of pounds per month of PET bottles. Dastou knows one PVC bottle among thousands of PET bottles can screw up an entire load.

PVC mixed with PET makes the batch too runny in manufacturing new PET products such as ski jacket filling, carpet, and geotextiles or road building materials.

Dastou said manual separating among thousands of bottles is too costly. Wellman currently has several proprietary ideas working to develop efficient mechanized processing, he said.

Although many in the recycling industry herald Society of Plastic Industries' (SPI) labeling on containers and bottles as a good start to efficient separation of plastics, the labeling is not complete for some post-consumer high density polyethylene (HDPE) end-users. HDPE is a number 2 in the plastic labeling system, but that number does not distinguish between blow-mold and injection-mold grades of HDPE.

And it should, according to recyclers like Gary Fish of Midwest Plastics whose HDPE piping made from remelted blow-molding grade HDPE is wrought with stress fractures if injection-mold grade HDPE is mixed into the batch. At 360 degrees Fahrenheit, blow-mold grade is like taffy, and injection-mold grade is closer to liquid form.

Blow-mold grades include oil bottles, milk jugs, and detergent bottles. Injection grade materials include cheese tubs, butter tubs, and yogurt containers.

The American Society for Testing and Materials is currently revamping the original labeling system designed by SPI. Fish and others hope the new system will end the blemished batches.

One plastic engineer, Morgan Gibbs of Quantum Chemical Corporation (Cincinnati, OH) has found that better quality control puts post-consumer HDPE's properties closer to virgin HDPE's in nonfood applications like oil and detergent bottles.

"Quality is going to be the key to future recycling," Gibbs said. Processors are advised to wash the plastic until all the labels and dirt are off the material, and manufacturers are to back up washing with several screens packed into their extruders. The screens will filter out any remaining impurities.

Environmental stress crack resistance (ESCR) and swell variation are two areas improved by a vigilance on quality, Gibbs said. ESCR tests a plastic's ability to withstand chemical attack, such as telling how the inner layer of a bleach bottle will fare, spending most of its product life next to a caustic liquid.

Swell variation examines how consistently the resin behaves during processing, which is important for reliable predictions such as telling whether or not a particular resin always melts at the same temperature.

Gibbs has shown in-house that quality procedures close the response gap for these characteristics between virgin HDPE and recycled HDPE. Future tests will deal with unsolved problems like odor.

Blow-molded containers are often permeated by the products they package, and the odor of these materials remains in the recycled resin. The contaminants also cause smoke during processing, Gibbs said. He will see if sandwiching post-consumer stuff between virgin layers eliminates this problem.

On the way to producing higher quality paper, glass, plastics, and aluminum recyclables, understanding requirements and inspection made by end-users is the first step. Contrary to popular belief, quality requirements are not designed to cause recyclers grief.

As new technologies change, quality specifications may also change. Communication between parties is crucial. Processors must take a large amount of responsibility for clean material. As a middleman between citizen and end-user, they must not only educate, but also ensure clean material. Recycling markets depend on it.

2 RECYCLING AND THE STATE AND FEDERAL GOVERNMENTS

As of press date, there is no U.S. federal recycling mandate, as there are in other countries (see Chapter 8), although in October 1993 President Clinton signed an executive order requiring federal agencies to purchase printing and writing paper with at least 20% post-consumer content by the end of 1994. States set recycling goals and other activities related to recycling. Understanding the goals of each state and the way they compute recycling rates and how they have or are legislating recycling is essential to understanding recycling in the U.S. The first two articles in this chapter describe the latest recycling rates in the United States and state legislation on recycling. The final article discusses the possibility of a National Recycling Bill.

STATE RECYCLING RATES

By *Lisa Rabasca,** editor of *Recycling Times*, Washington, D.C.

Recycling rates reported by states in 1993 appear to have hit a plateau. Fewer than 20 states reported increases in their recycling rates, according to the 1993 annual *Waste Age's Recycling Times* survey on how states quantify recycling rates and count municipal solid waste (MSW).

Nearly half the state recycling rates remain in the 5–15% range, with only a handful of states reporting rates above 25%. **Maryland** and **Ohio** experienced the largest growth in their recycling rates.

---

*Susan Edquist, Jennifer Goff, Nancy Lang, Michael Malloy, Randy Woods, and Kathleen White also contributed to this report.

A stagnant growth in recycling rates may be a cause of concern for many states as 1995 approaches. In the late 1980s and early 1990s, many state legislatures set 1995 as a benchmark year for meeting recycling and waste reduction goals. Only a handful of states are close to reaching their goals, survey results reveal. However, some states say they have not calculated a new recycling rate since 1992 or 1993.

After several years of *Waste Age's Recycling Times* surveys, two facts remain: Many states still guess or estimate their recycling rates, and there are myriad ways to count recycling rates and MSW generation.

When determining recycling rates, most states include composting, but few states include source reduction. Ten states include some form of incineration as either recycling or waste reduction. See Table 2.1 for a detailed list of what each counts in its recycling rate, as well as the state's recycling rate and MSW generation.

Although several state officials say there should be a standardized method among states for counting recycling rates and MSW generation, few state officials seem to have any solid suggestions for achieving continuity. Many state officials say a standardized method of counting recycling rates is unrealistic, given that states have different recycling laws.

Once recyclables are collected, they must be marketed and sold to have any effect on waste reduction. The survey reveals that few states are able to track whether collected recyclables are actually marketed and sold.

New Ways to Count

Two states—**California** and **Massachusetts**—have changed the way they count recycling rates and MSW generation.

As of 1993, **California** changed from generation-based counting to disposal-based counting, using the 1990 base-year generation amounts to determine the amount of diversion needed to meet the state's goals of 25% waste diversion by 1995 and 50% by 2000. Currently, California reports an 11% recycling rate. However, this figure has not been updated since 1990, according to state officials.

In a disposal-based system, the number of tons disposed at permitted solid waste facilities is compared to 1990 base-year generation data, explains Tom Estes, [then-] spokesman for the California Integrated Waste Management Board (IWMB, Sacramento, CA). Each jurisdiction estimates the amount of solid waste generated in any given year by adjusting its base-year generation amount for changes in population, economics, and other factors. The adjustment method is being developed, Estes says, and factors being considered include population, employment, taxable transactions, and IWMB-approved disaster-related wastes.

The amount disposed will be subtracted from the amount generated, and any diversion program that reduces the amount of disposal will count toward diversion. Jurisdictions will only be required to report on recycling and composting programs they fund or operate, Estes explains. Jurisdictions will not need to quantify source reduction, private sector recycling, and private sector composting, he adds.

Massachusetts expanded its survey for estimating residential waste generation. "We got numbers on the total that was disposed and diverted, commercial generation, and subtracted residential," explains Jeffrey Lissack of Massachusetts Department of Environmental Protection. "What you're left with is commercial. We also looked at commercial processors, and we came up with tons per employee and extrapolated off of that."

While Massachusetts reports an increase in its recycling rate to 26% from 22%, Lissack cautions that it may not be an actual increase, but the result of having better data.

Making Compost Count

There is no clear consensus on whether to count composting, source reduction, and incineration as recycling. While many of the states count composting in their recycling rate, several states offer caveats for counting composting.

While **Illinois** does not count composting in its recycling rate, yard waste taken to a compost facility is counted separately. In 1993, yard waste represented 3% of the state's total MSW stream, state officials say. Backyard composting is not counted by the state at all.

In **Indiana**, districts may claim up to 17% of their MSW generation as composting, according to U.S. Environmental Protection Agency (EPA) estimates.

Pennsylvania only counts composted yard waste in its recycling rate, **South Dakota** counts composting as "re-use" rather than recycling, and **Vermont** considers composting as source reduction.

Wyoming estimates how much it composts, based on population, how much waste is generated, and how many municipalities are composting.

Table 2.1 What Each State Counts as Its Recycling Rate

	MSW Recycling (million tpy)	Recycling/Waste Reduction Goal/yr	Estimated Recycling Rate[b]	How Did the Rate Change?	Packaging?	Household Durable Goods	Yard Waste	Solid Waste Composting
Alabama	5.1	25%	10%	+	X	X	X	X
Alaska	NA	NG	NA					
Arizona	4	NG	5–6%	same		X		X
Arkansas	1.9	40%–2000	10%	+		X	X	X
California	44.5	50%–2000	11%		X	X	X	X
Colorado	4.4	50%–2000	NA					
Connecticut	2.9	50%–2000	21%	+	X		X	
Delaware	0.8	NG	28%	–		X		
Dist. of Columbia	0.9	45%–1996	30%	same	X	X	X	X
Florida	20	30%–1994	30%	+	X	X	X	X
Georgia	8	25%–1996	NA					
Hawaii (Honolulu only)	1.3	50%–2000	12%	same	X	X	X	X
Idaho	0.9	25%–1995	8–10%	same	X	X	X	
Illinois	14.1	25%–2000	17%	+	X	X		
Indiana	3.9	50%–2001	NA		X	X	X	X
Iowa	2	50%–2000	16%	+	X	X	X	X
Kansas	NA	NG	NA					
Kentucky	3.2	25%–1997	9–11%	+			X	X
Louisiana	3.48	25%–1992	5–10%	same	X	X	X	X
Maine	1.2	50%–1994	NA		X	X	X	X
Maryland	5	20%–1994	25%	+	X	X	X	X
Massachusetts	6.4	56%–2000	26%	+	X	X	X	X
Michigan	13.5	45%–2005	NA					
Minnesota	4.4	25%–1995	39%	+	X	X	X	X
Mississippi	2.3	25%–1996	5%	+	X	X	X	X
Missouri	4.8	40%–1998	NA					
Montana	0.7	25%–1996	5–7%	+	X	X	X	X
Nebraska	1	50%–2002	16%	+	X	X		
Nevada	3.6	25%–1995	12%	+	X	X	X	
New Hampshire	1	40%–2000	7–8%	same			X	
New Jersey	14.3	60%–1996	52%	same	X	X	X	X
New Mexico	1.5	50%–2000	6%	same	X		X	
New York	22.8	50%–1997	11–23%	same	X		X	X
North Carolina	6.8	40%–2001	6%	same		X	X	X
North Dakota	0.5	40%	12–15%	same	X	X	X	X
Ohio	17.4	25%–1994	23%	+	X	X		
Oklahoma	3	NG	NA					
Oregon	2.6	50%–2000	27%	+	X		X	
Pennsylvania	1.2	25%–1997	16%	+	X		X	
Rhode Island	0.6	70%	14%	same				
South Carolina	5	30%–1997	6–8%	+			X	X
South Dakota	0.7	50%–2001	5–8%	same	X	X	X	X
Tennessee	5.4	25%–1995	8%	+	X	X	X	X
Texas	18–20	40%–1994	12%	+	X	X	X	
Utah	1.5	NG	15%	same		X	X	X
Vermont	NA	40%–2000	25%	same		X	X	
Virginia	7.6	25%–1995	19%	same		X	X	
Washington	6	50%–1995	35%	+	X	X	X	
West Virginia	1.5	50%–2010	13–14%	+		X	X	
Wisconsin	NA	[a]	NA					
Wyoming	0.3	NG	3%		X	X	X	

[a]Wisconsin's law does not set a specific recycling goal. Instead, it bans disposal of most recyclable and compostable materials and requires "effective recycling programs" at the local level. Program approval creates an exemption from the ban.
[b]States differ as to what materials count when assessing recycling rates.

Table 2.1 What Each State Counts as Its Recycling Rate

Automobiles	Metal Scrap	Post-Industrial Plastic Scrap	C&D Waste	Tires	Batteries	Used Oil	Other HHW	Source Reduction	Commercial	Incineration	Other	Could Not Specify
X	X	X	X	X	X	X	X	X	X	X	X	
												X
X									X			
				X	X	X						
X	X	X	X	X	X	X	X	X	X	X	X	
												X
	X				X	X						
X	X	X	X	X	X	X				X		
X	X	X	X	X	X				X			
	X	X	X	X	X	X	X		X			
												X
X	X	X	X	X	X	X	X		X			
	X											
	X											
	X	X	X	X	X	X	X	X	X			
X		X	X	X	X	X	X	X	X		X	
												X
			X	X								
X	X	X	X	X	X	X	X	X	X	X	X	
			X	X	X		X		X			
	X		X	X					X	X		
	X								X			
												X
		X							X	X		
X	X	X	X	X	X	X	X		X			
												X
					X						X	
				X								
X	X	X	X	X			X		X		X	
	X		X	X	X	X		X				
X	X	X	X	X	X	X	X		X			
	X		X	X	X	X			X			
	X				X	X	X	X				
X	X	X	X	X	X	X	X	X	X	X	X	
X	X	X		X	X	X		X			X	
												X
				X		X	X		X			
		X							X			
												X
	X	X	X			X	X		X			
	X			X	X	X	X	X	X			
X	X	X	X	X	X	X	X	X	X			
X	X	X	X				X	X	X			
		X							X		X	
X	X	X	X	X	X	X			X			
X	X	X		X	X	X			X			X
			X									
												X
X	X			X	X	X			X			

KEY: NA = Data not available; NG = no goal; tpy = tons per year.
Source: *Waste Age's Recycling Times*, March 1994, pages 62–63.

Source Reduction as Recycling

Few states include source reduction efforts in their recycling rates, and, for the most part, those that do, lack a formal method for measuring source reduction efforts. However, this is not surprising because source reduction is difficult to quantify.

Arizona and **Idaho** count source reduction as pollution prevention prior to recycling, and **Vermont** counts source reduction as diversion rather than recycling.

Iowa measures source reduction efforts by using a disposal-based quantification methodology. Landfill tonnages and incineration amounts for a given year are compared with data from 1988 on a per capita basis. Any diversion is assumed to be the result of source reduction and recycling activities, explains Tom Collins of the Iowa Office of Solid Waste.

Minnesota has begun a program to credit counties for their source reduction activities. Beginning in 1995, counties with an approved source reduction program can get up to a 3% credit toward their recycling goals, explains Tony Hainault, policy analyst with the Minnesota Office of Waste Management.

Is Incineration Recycling?

Ten states allow some form of incineration to be counted as recycling or waste reduction.

In **California**, up to 10% of the 50% disposal reduction mandate may be incineration. Refuse-derived fuel is counted as recycling in **Delaware**.

Georgia has one incinerator that burns MSW. Generally, the state does not include incineration in its waste reduction rate. However, if a company began burning MSW (which had previously been landfilled) to generate energy, this would apply toward the state's waste reduction rate.

Idaho allows incineration and counts it as waste diverted from the landfill, rather than recycling. Two waste-to-energy (WTE) facilities that were operating in **Maryland** prior to 1988 receive a 5% credit toward their recycling goals. Similarly, **Oklahoma** allows some WTE facilities to qualify as recycling. Incineration is included in **Rhode Island** as resource recovery. In **Kentucky**, only the pyrolysis of tires counts as recycling.

Finding Common Ground

State officials agree that finding common ground in how to count recycling rates is not an easy task. While many state officials admit they are comparing "apples to oranges," most state officials say they would not want the EPA dictating how recycling rates should be determined. Several groups, including the EPA, the Council of Great Lakes Governors (Chicago), and the Council of State Governors (Lexington, KY) are trying to foster standardized recycling rates.

"It is very difficult to compare programs across the states when what is counted in a recycling rate is so different," says John Hendren, staff economist at the Illinois Department of Energy and Natural Resources' Office of Recycling and Waste Reduction.

"I do feel there are a lot of states out there using bad data (guesses) to come up with recycling rates that really have no validity out of [the] need to measure success against legislative goals," says Sharon Edgar of the Michigan Department of Natural Resources. Michigan does not estimate a recycling rate because state officials realize they do not have the data to support a decent estimate, Edgar notes.

Wisconsin has an approach similar to Michigan's: it has not set a recycling rate goal. "Recycling rate calculations quickly turn into a numbers game, and one can never be sure if any two recycling rates are comparable," says Kate Cooper, section chief of the Waste Reduction and Recycling Program, Bureau of Solid and Hazardous Waste at the Wisconsin Department of Natural Resources.

"Standardization in and of itself is less critical than standardization that facilitates better waste management programs across the states and meets industry needs for information for market development," says Minnesota's Hainault.

Sometimes even within a state there is dissension on how to count recyclables. In **New York**, the state Department of Environmental Conservation reports the state recycling rate at 23%, while the Legislative Commission on Solid Waste reports an 11% state recycling rate.

Because states have different laws, some officials say a standardized method for states to count their recycling rates is not realistic.

"Since state laws are different and have different goals, it will be difficult to standardize recycling rate calculations," says Estes of California's IWMB.

"I don't think it will ever happen," said Ron Hendricks, environmental administrator for the Waste Reduction Section of Florida's

Department of Environmental Protection. "It will be easier to compare what states do and don't recycle rather than coming up with a standardized method."

"Guidelines would be helpful, but to change all factors in recycling equations would not be cost beneficial to any state," says Michele Retell, public affairs director for Virginia's Department of Environmental Quality.

Selling What They Collect

States are equally unsure about how much of what they collect for recycling is actually sold and marketed to end-users. While most states admit they do not track such information, several say they assume everything collected is sold. A few states admit they guess or go by rumors or word of mouth.

Delaware claims 69,000 tons of recyclables were sold last year. This figure is based on reports from Delaware's Solid Waste Authority and recycling companies across the state.

More than 90% of what is collected is sold, **Hawaii** claims. The state determines this figure by comparing reports from different collection sources.

Illinois admits it does not know how many recyclables are sold; however, state officials have unsuccessfully tried to pass legislation that would require measurement and reporting of recyclables sold in the state.

An average of 2.5% to 4% of the recyclables collected in **Iowa** are not sold because of contamination problems, according to state officials.

Kansas claims that 25 of the state's largest recycling facilities use 50% of the recyclables collected in the state.

"We don't collect [recyclables] without a market," says Charles Peters of **Kentucky's** Division of Waste Management. However, Peters admits the state has no statistical information on whether all recyclables collected are sold.

Maine measures how many recyclables are sold by comparing reports from municipalities, surveys from the commercial sector, and data from the state's bottle deposit program. State officials say almost all of what is collected is sold.

Massachusetts claims to be an importer of newspaper and glass. "We've done surveying and came up with a number that is greater than the tons collected," says Lissack.

Based on reports by municipalities, **New Hampshire** claims it sold 59,700 tons of recyclables in 1992.

Washington's annual survey determined that 2.15 million tons of recyclables were sold last year.

****

RECYCLING IN THE STATES - 1993 UPDATE

By *Chaz Miller*, recycling manager for the Environmental Industry Associations, Washington, D.C.

1993, like 1992, was a slow year for recycling legislation. As in 1992, most states only tinkered with previously enacted laws. On the national level, Congress didn't even tinker and no new recycling laws were passed.

Most state action in 1993 focused on market development. While markets remain weak, says Michelle Raymond, editor of *State Recycling Laws Update* (Riverdale, MD), "none of the states have repealed their existing recycling rate mandates, despite the high cost of recycling."

At the White House

Recycling's biggest boost in 1993 came from President Clinton's Executive Order establishing recycled content levels for federal purchases of printing and writing paper.

Released on October 20, the order directs every federal agency to purchase printing and writing paper containing 20% post-consumer materials by the end of 1994 and 30% post-consumer material by the end of 1998. The order allows sawdust to be counted as post-consumer material.

President Clinton also required federal agencies to use re-refined oil and to replace virgin tires with retreads. Perhaps most importantly, the order requires federal agencies to revise their procurement specifications and standards so that recovered materials can be used to make federally bought products.

Ironically, the North American Free Trade Agreement (NAFTA), which President Clinton fought so hard for in 1993, might undo the

Executive Order. Language in the NAFTA procurement section could be used to overrule the post-consumer content requirements as a barrier to free trade. However, states are not yet required to adopt NAFTA's procurement standards and may continue their own aggressive recycled content procurement programs.

President Clinton also endorsed volume-based fees (also known as "pay by the bag") for garbage collection in the Climate Change Action Plan. Released one day before the procurement Executive Order, the plan proposes spending $11 million in FY 1995 on source reduction and recycling activities.

On the state level, continued problems with recycling markets forced states to concentrate on demand-side, market development legislation. States did not expand comprehensive collection programs. Volume-based fee systems are only required by three states. In addition, several East Coast cities and counties stopped volume-based fee programs after taxpayer protests.

On the Supply Side

Statewide supply-side legislation in 1993 included:

- Thirty-nine states and the District of Columbia have some form of statewide recycling law (see Table 2.2).
- Seven states and the District of Columbia require local government to mandate source separation of one or more recyclables from solid waste (Maine's law applies only to office paper and corrugated containers from businesses). Some of these states limit mandatory recycling to larger towns in the state.
- Twenty states require that local government prepare recycling plans in order to meet waste reduction or recycling goals. None of these laws explicitly require recycling. Nonetheless, some of the most active state recycling programs, such as California's, are in states that only require recycling plans at the local level.
- Twelve states require more than plans but less than mandated separation. In these states, local government must ensure that recycling opportunities are available. Curbside collection, drop-off centers, or mechanical processing of mixed waste are potential options.

Connecticut and North Carolina increased their recycling goals from 25% to 40%. Rhode Island rejected attempts to revise its 1992 law banning incineration and set a statewide goal of processing 70% of the waste stream for recycling.

Table 2.2 Types of Recycling Plans or Laws

Recycling plans = a requirement for local government to prepare a recycling plan. In many cases, the plan is tied to a recycling/waste reduction goal. However, the state law does not explicitly require that local government operate a recycling program.

Opportunity to recycle = a requirement that local government offer the opportunity to recycle through curbside collection, or drop-off center, or by processing recyclables from mixed waste.

Source separation = a requirement that local government mandate the source separation of one or more recyclables.

ALABAMA - Recycling plans
ARKANSAS - Opportunity to recycle
ARIZONA - Opportunity to recycle
CALIFORNIA - Recycling plans
CONNECTICUT - Source separation
DELAWARE - Opportunity to recycle
DISTRICT OF COLUMBIA - Source separation
FLORIDA - Opportunity to recycle
GEORGIA - Opportunity to recycle
HAWAII - Recycling plans
ILLINOIS - Recycling plans
INDIANA - Recycling plans
IOWA - Recycling plans
KENTUCKY - Recycling plans
MAINE - Source separation for businesses (office paper, corrugated), statewide plan
MARYLAND - Recycling plans
MASSACHUSETTS - Recycling plans
MICHIGAN - Recycling plans
MINNESOTA - Opportunity to recycle
MISSOURI - Recycling plans
NEVADA - Opportunity to recycle (urban), recycling plans (rural)
MEW HAMPSHIRE - Recycling plans
NEW JERSEY - Source separation
NEW MEXICO - Recycling plans
NEW YORK - Source separation
NORTH CAROLINA - Opportunity to recycle
OHIO - Recycling plans
OKLAHOMA - State law requires county plans
OREGON - Opportunity to recycle
PENNSYLVANIA - Source separation
RHODE ISLAND - Source separation
SOUTH CAROLINA - Opportunity to recycle
SOUTH DAKOTA - Recycling plans
TENNESSEE - Recycling plans
TEXAS - Recycling plans
VERMONT - Recycling plans
VIRGINIA - Recycling plans
WASHINGTON - Opportunity to recycle
WEST VIRGINIA - Source separation (urban only)
WISCONSIN - Opportunity to recycle

Table 2.3 Disposal Bans

	Lead Acid Batt.	Yard Waste	Whole Tires	Used Oil	White Goods	Other
Alabama	X			X		
Alaska				X		
Arizona	X		X	X		
Arkansas	X	X	X	X		
California	X		X	X	X	A,B,U
Colorado						
Connecticut	X	P	X	X	X	S
Delaware						
Florida	X	X	X	X	X	C,G,S
Georgia	X	X	X	X		
Hawaii	X				X	
Idaho	X		X			
Illinois	X	X	X	X	X	
Indiana	X	X	X			
Iowa	X	X	X	X		D,S
Kansas			X	X		
Kentucky	X		X	X		
Louisiana	X		X			
Maine	X		X	X	X	S
Maryland		P	X			S
Massachusetts	X	X	X		X	D,E
Michigan	X	X		X		
Minnesota	X	X	X	X		F,G,H,Q,S
Mississippi			X	X		
Missouri	X	X	X	X	X	
Montana						
Nebraska	X	X	X	X	X	I
Nevada				X		
New Hampshire	X	X	X	X		
New Jersey	X	X	X	X	X	B,S
New Mexico	X					
New York	X	P	X	X		
North Carolina	X	X	X		X	J,R
North Dakota	X			X	X	
Ohio	X	X	X			
Oklahoma	X					
Oregon	X	P	X	X	X	K,S
Pennsylvania	X	X		X		
Rhode Island		X		X		L,M,S
South Carolina	X	X	X	X	X	
South Dakota	X	X	X	X	X	N

Table 2.3 Disposal Bans (Continued)

	Lead Acid Batt.	Yard Waste	Whole Tires	Used Oil	White Goods	Other
Tennessee	X		X			
Texas	X		X	X		
Utah	X		X	X		
Vermont	X		X	X	X	S
Virginia	X	P		X		
Washington	X					
West Virginia	X	X	X	X		
Wisconsin	X	X	X	X	X	O,T
Wyoming	X			X		

A–generators of more than 25 fluorescent lights
B–scrap metal
C–construction and demolition debris
D–deposit containers
E–recyclable paper, single polymer plastics
F–dry cell batteries with mercuric oxide or silver oxide electrodes
G–mercury-containing products such as batteries and fluorescent lights
H–telephone directories
I–some small quantity commercial and household hazardous wastes
J–wet cell batteries
K–separated recyclable material
L–loads of commercial waste with more than 20% recyclables
M–mercury oxide batteries
N–glass, metal, and rigid plastic containers; corrugated boxes; other forms of paper packaging
O–nondegradable yard waste bags; aluminum, plastic, steel, and glass containers; cardboard, foam polystyrene packaging; magazines, newspapers, and office paper unless municipalities are certified as having an "effective" source separation program
P–these states promote yard and/or leaf waste composting by allowing local jurisdictions to ban disposal or by requiring composting if it is economically feasible, but they stop short of a ban on disposal
Q–brake fluid, power steering fluid, transmission fluid, motor oil filters, and antifreeze.
R–antifreeze, aluminum and steel cans
S–rechargeable batteries
T–fluorescent lights
U–household hazardous waste

Check with the state for details about date of implementation.
SOURCES: *Waste Age*, State Recycling Laws Update, State Legislation.

Forty-two states have official recycling or waste reduction goals (see Table 2.1). Maryland's 20% recycling goal is at the low end, while Rhode Island's 70% processing for recycling goal is at the high end.

States seem to be tending toward waste reduction goals. Recycling goals require a bureaucracy to track amounts of materials collected at the city and county level. They also need an elaborate system to eliminate doublecounting among collectors, processors, and end markets. Waste reduction goals simply require population and waste disposal data. These goals also have the advantage of emphasizing source reduction.

Disposal Bans

1993 saw a continuation of 1992's trend of passing fewer disposal bans (Table 2.3). While disposal bans declare where particular materials *cannot* be placed, they do not say where they can be placed. As a result, bans are becoming less popular as a quick fix. Arkansas delayed its yard waste disposal ban and West Virginia delayed yard waste, scrap tire, and lead-acid battery disposal bans.

In 1993, Connecticut expanded its restrictions on yard waste disposal, while Georgia banned yard waste from disposal. Florida banned fluorescent lights, most mercuric oxide batteries, and rechargeable batteries from disposal. Minnesota banned the disposal of most types of automotive fluids and motor oil filters. North Carolina banned landfilling of steel cans and white goods, and landfilling and incineration of antifreeze and aluminum cans. Oklahoma enacted a ban on disposal of lead-acid batteries. Utah banned disposal of whole tires.

Several states banned the sale of mercury-containing batteries. This type of legislative action is more direct than a disposal ban. In addition, 10 states now effectively prohibit from disposal in solid waste some or all rechargeable batteries.

In a related development, Florida repealed its degradability requirements for plastic bags.

On the Demand Side

1993 recycled-content legislation included:

- Only three states adopted recycled-content requirements in 1993. Colorado and Hawaii signed voluntary newspaper recycled-content agreements, while Florida legislated specific recycled-content requirements for newsprint.

- Thirteen states now have recycled-content legislation for newsprint. Six of these states also require recycled content in other products such as glass and plastic containers and telephone books (see Table 2.4). An additional 14 states have voluntary recycled-content agreements covering newspapers. These 27 states have more than 70% of the American newsprint market.
- In both California and Oregon, the plastics industry fought vigorously to delay or to eliminate recycled-content requirements for glass and plastic containers. In both states, the issues were similar. On the one hand, industry representatives argued that they have little control over the recycling rate and that recycled plastic could not be used for food contact packages. Recyclers and environmentalists pointed out that recycled plastic can be used for some food products and that the requirements are essential for market development. In Oregon, the plastics industry urged that pyrolysis of plastics be counted as recycling.
- California's Senate approved a bill exempting plastic food, drug, and cosmetic containers from recycled content requirements. The Senate bill also required that 60% of California households have curbside collection of all rigid plastics before any recycled-content requirements would go into effect. Although the Senate bill died in a House Committee, the deadline for recycled content in plastic food packaging was delayed for two years.
- In Oregon the plastics industry delayed passage of the state environmental office's budget until its concerns over Oregon's plastic minimum-content law were addressed. The final resolution included a one-year delay, until 1996, in enforcing the plastic recycling provisions of the Oregon law. No decision was reached on pyrolysis. Instead, the issue of counting pyrolysis as recycling will be studied.
- At the end of the year, California's Integrated Waste Management Board adopted a "utilization for recycling goal" for 50% of the post-consumer wastepaper generated in California.

A growing number of states are providing tax credits to support the use of recycled materials. North Dakota established sales and use tax exemptions for recycling machinery. Montana doubled its 5% income tax deduction for purchase of business-related materials made from recycled content. Nevada now allows property tax exemptions and electricity discounts for recycling businesses. Hawaii lowered the

Table 2.4 Minimum Content Laws

ARIZONA:	Newsprint: 50% by 2000
CALIFORNIA:	Newsprint: 50% by 2000
	Glass containers: 65% by 2005
	Plastic bags: 10% for 1.0 mil bags, 1993
	30% for .75 mil bags, 1995
	Plastic containers: similar to Oregon law
	Fiberglass: container cullet where feasible
CONNECTICUT:	Newsprint: 50% by 2000
	Telephone directories: 40% by 2001
FLORIDA:	Newsprint: 30% by 1995
ILLINOIS:	Newsprint: 23% by 1993
MARYLAND:	Newsprint: 40% by 1998
	Telephone directories: 40% by 2000
MISSOURI:	Newsprint: 50% by 2000
NORTH CAROLINA:	Newsprint: 40% by 1997
OREGON:	Glass containers: 50% by 2000
	Newsprint: 7.5% by 1995
	Plastic containers: 25% post-consumer material or 25% recycling rate or reusable/refillable 5 times or reduced 10% in content by 1995 (certain exemptions)
RHODE ISLAND:	Newsprint: 40% by 2001
TEXAS:	Newsprint: 30% by 2000
WEST VIRGINIA:	Newsprint: "highest practicable content," advisory committee to determine rate
WISCONSIN:	Newsprint: 45% by 2001
	Plastic containers: 10% by 1995
Voluntary Agreements:	
COLORADO:	Newsprint: 30% by 1998
HAWAII:	Newsprint: "greatest extent possible"
INDIANA:	Newsprint: 40% by 2000
IOWA:	Newsprint: 40% by 2000
LOUISIANA:	Newsprint: 40% by 2000
MAINE:	Newsprint: 40% by 1995
MASSACHUSETTS:	Newsprint: 50% by 2000
MICHIGAN:	Newsprint: 24% by 1995
NEW HAMPSHIRE:	Newsprint: 40% by 2000
NEW YORK:	Newsprint: 40% by 2000
	Telephone Directories: 40% by 1998
OHIO:	Newsprint: 40% by 2000
PENNSYLVANIA:	Newsprint: 50% by 1995
VERMONT:	Newsprint: 40% by 2000
VIRGINIA:	Newsprint: 30% by 1995

NOTE: Requirements concerning type of secondary fiber, and recycled newsprint as percentage of newsprint purchases and/or as percentage of total fiber used vary among state laws and agreements.

tax rate on recycling industries. Several other states expanded, reapproved, or made technical corrections to existing tax credits.

Twenty-eight states now have tax incentives for recycling. These incentives vary dramatically as to extent, scope, and applicability (see Table 2.5).

State Market Development

At least nine states either created recycling market development councils in 1993 or required recycling market studies. Sixteen states now have active Market Development Councils. Interestingly, 11 of these Councils are in southern states. They are popular in the south because "they can serve as technical advisors to state legislators and help businesses overcome barriers to recycling markets," says Steve Moye of the Southern Legislative Conference (Atlanta, GA).

Washington's Clean Washington Center (CWC), a Division of the Washington Department of Trade and Economic Development, received a $1.2-million federal grant to create the Recycling Technology Assistance Project. CWC will work with local manufacturers to encourage the substitution of recycled materials for virgin feedstock. The National Recycling Coalition (Washington, DC), will be involved with CWC in setting up a clearinghouse for sharing recycling technology.

Procurement

Every state has legislation or a policy encouraging the procurement of recycled products with secondary content by state agencies and contractors. While the majority of these laws are limited to paper products, they increasingly cover other products such as compost and motor oil.

Thirty-eight states apply a price preference to paper with secondary content. The price preference is usually 5% to 10% above the price of competing virgin paper. Thirty-one states "set-aside" specific percentages of paper purchases for paper with secondary content, and 22 states have both set-asides and price preferences (see Table 2.6). Six states limit their procurement policy to a general declaration that buying products with secondary content is a "good thing."

Nine states expanded their procurement policies in 1993. Actions ranged from California and Colorado requiring all legal documents to use recycled-content paper to Florida, North Carolina, and New Jersey

Table 2.5 Recycling Tax Credits

ARIZONA: Income tax credit of 10% of installed cost of recycling equipment up to lesser of 25% of total tax liability or $5,000.

ARKANSAS: 30% tax credit on income/corporate taxes for purchase of equipment to make products with at least 10% recycled content.

CALIFORNIA: Banks and corporations may take a 40% tax credit for the cost of equipment used to manufacture recycled products with minimum 50% secondary content and 10% post-consumer content. Development bonds are available for manufacturing products with recycled materials.

COLORADO: Up to 20% tax credit for purchase of certain equipment to make products using post-consumer recycled materials. Special credits for plastic recycling.

DELAWARE: Corporate tax credits for investments and for job creation for use of minimum 25% secondary materials removed from in-state waste stream. Reductions in gross receipts tax also apply. Corporate tax credits also available for source-reduction activities and for processors and collectors of recyclable materials.

FLORIDA: Tax incentives to encourage affordable transportation of recycled goods from collection points to sites for processing and disposal.

HAWAII: Defines recycling as a manufacturing, not a service industry, thus allowing a lower tax rate for recycling businesses.

ILLINOIS: Sales tax exemption for manufacturing equipment.

INDIANA: Property tax exemption for buildings, equipment, and land involved in converting waste into new products.

IOWA: Sales tax exemptions for recycling equipment.

KANSAS: Tax abatement for equipment used to manufacture products made with at least 25% post-consumer material.

KENTUCKY: Property and income tax credits to encourage recycling industries.

LOUISIANA: Corporation and franchise tax credits for purchase of qualified recycling equipment; corporate and personal income tax credits for purchase of equipment to recycle CFCs used as refrigerants.

MAINE: Corporate tax credits equal to 30% of cost of recycling equipment and machinery. Tax credits of up to $5 per ton of wood waste from lumber products used as fuel or to generate heat.

MARYLAND: Individual and corporate income credit for expenses incurred to convert a furnace to burn used oil or to buy and install equipment to recycle used Freon.

MINNESOTA: Sales tax exemptions for recycling equipment.

Table 2.5 Recycling Tax Credits (Continued)

MONTANA: 25% tax credit on purchase of equipment to process recyclable materials; up to 10% off income taxes for purchase of business-related products made with recycled material.

NEVADA: Personal property exemptions for new businesses that manufacture products using recyclables and electricity discounts through public utilities for recycling businesses.

NEW JERSEY: 50% investment tax credit for recycling vehicles and machinery. 6% sales tax exemption on purchases of recycling equipment.

NEW MEXICO: Tax credits on equipment to recycle or use recycled materials in a manufacturing process.

NORTH CAROLINA: Industrial and corporate income tax credits and exemptions for equipment and facilities.

NORTH DAKOTA: Sales and use tax exemptions for recycling machinery and equipment.

OKLAHOMA: 15% income tax credit on purchase of equipment and facilities to use recyclable materials in a product.

OREGON: Individual and corporate income tax credits for capital investment in recycling equipment and facilities. Special credits for plastic recycling.

SOUTH CAROLINA: Scrap metals dealers defined as manufacturers for sales tax purposes and exempted from electricity and fuel sales taxes.

TEXAS: Sales tax credits for businesses. Sludge recycling corporations eligible for franchise tax exemptions.

VIRGINIA: Individual and corporate tax credits of 20% of the purchase price of machinery and equipment for processing recyclable materials. Manufacturing plants using recycled products are eligible for a 10% tax credit.

WASHINGTON: Motor vehicles are exempt from rate regulation when transporting recovered materials from collection to reprocessing facilities and manufacturers.

WEST VIRGINIA: Disposal tax waivers for commercial recyclers who reduce their solid waste by 50%.

WISCONSIN: Sales tax exemption for waste reduction and recycling equipment and facilities; business property tax exemptions for same equipment.

In addition, many states have tax credits that apply to new business in general, to business expansions, or to business locating in certain pre-designated areas.

In all cases, consult with state Commerce, Economic Development, or Tax Offices to learn the details of state law.

Table 2.6 State Procurement Laws for Paper Products

	Price Preference	Set Aside
Alaska	X	X
Arizona	X	
Arkansas	X	X
California	X	X
Colorado	X	X
Connecticut	X	
Delaware		X
District of Columbia	X	X
Florida	X	X
Georgia	X	
Illinois		X
Indiana	X	
Iowa	X	X
Kansas	X	X
Louisiana	X	X
Maine	X	X
Maryland	X	X
Massachusetts	X	
Michigan	X	X
Minnesota	X	
Mississippi	X	
Missouri	X	X
Montana		X
Nevada	X	
New Hampshire	X	
New Jersey	X	X
New Mexico	X	
New York	X	
North Carolina		X
North Dakota		X
Oklahoma	X	X
Oregon	X	X
Pennsylvania	X	
Rhode Island	X	
South Carolina	X	X
South Dakota	X	X
Tennessee		X
Texas	X	
Utah	X	X
Vermont	X	X
Virginia	X	
Washington	X	X
West Virginia	X	X
Wisconsin		X
Wyoming		X

SOURCE: Richard Keller, Northeast Maryland Solid Waste Authority.

establishing 65% set-aside goals. "Major 1994 procurement activity will be state and local governments incorporating the federal Executive Order into their local procurement policies," predicts Richard Keller, an expert on state paper procurement requirements.

Other Legislation

In other recycling-related actions, Florida allowed recycling trucks to stop while collecting materials, but requires that they have a flashing amber light while stopped. Voters in Fairfax, CA, banned the use of polystyrene packaging. Finally, West Virginia banned incineration of municipal solid waste.

Advance Disposal Fees

The theory of advance disposal fees (ADF) is simple—require the cost of each product in the waste stream to include a fee for its disposal cost. Revenue raised by the fee will ensure that the material is properly managed. More than half the states charge disposal fees for tires. These fees are usually one or two dollars per tire. To date, tires have a disposal rate in excess of 77%, millions of tires are in piles around the country, and tire ADFs supply states with a steady source of revenue.

Nonetheless, ADFs are often advocated as a way to encourage recycling of packaging and other materials. In 1988, Florida passed America's first packaging ADF. Florida's law called for a one-cent-per-container ADF starting on October 1, 1992 and increasing to two cents per container in 1995. Containers recycled at a 50% rate did not have to pay the ADF.

Legislative bickering over ADF implementation delayed its effective date for one year. Manufacturers and packagers battled for provisions favorable to their material. Retailers and wholesalers fought over where the tax would be levied. At one point, a Florida grocery store chain threatened to stop selling a very popular beer because the beer's brewer supported charging the ADF at the wholesale level.

When the dust settled, the revised law required a one-cent-per-container fee to be assessed at the wholesale level on containers that had not achieved a 50% recycling rate. Fee assessments began on October 1, 1993. The ADF increases to two-cents-per-container in 1995. Aluminum and steel cans are recycled at a 50% rate in Florida and are not being assessed the ADF.

The ADF can also be removed for materials that meet recycled-content goals. These include a 35% goal for glass containers by mid-1994 and 50% by the end of 1997. Plastic containers must meet a 25% recycled content rate by mid-1994. Paper packaging, including drink boxes, must meet a 30% rate by mid-1994 and a 40% rate by the end of 1996. Glass and plastic can meet their requirements by being recycled into other products such as glasphalt or carpet. Paper packaging can avoid the ADF if it achieves a 30% recycling rate by mid-'94, 40% by mid-'95, and 50% by 2002.

Florida also increased the 10-cent-per-ton tax on virgin newsprint used in Florida to 50 cents per ton. If Florida newspapers are recycled at a 50% rate, the tax does not apply. It did not apply in 1993.

What Will Happen in 1994 and Beyond?

States will continue to focus on market development—supply-side legislation is more common during periods of economic strength. While the economy appears to be improving, most state legislatures will look to increase the demand for recyclables, not the supply. They will also be inclined to wait to see if previously enacted recycling or waste reduction goals are met.

****

IS A NATIONAL RECYCLING BILL POSSIBLE?

By *Jennifer A. Goff*, managing editor for *Recycling Times*, Washington, D.C.

Despite the proactive environmental positions of the Clinton administration and 103rd Congress, recycling has remained ancillary to other national issues such as universal health care, Superfund, the Safe Drinking Water and Clean Water Acts, and solid waste flow control.

Though some effort has been made on the federal level to address problems facing the recycling industry, comprehensive, national recycling legislation has failed to emerge [as of the 103rd Congress].

Still, the action that has been taken on the federal level for 1992–1994 may serve as both a model and a springboard that will spur the development of a national recycling bill. In fact, the road toward federal legislation is currently being paved by at least one national recycling organization.

NRC Moves Forward

For many in the recycling community, coming up with some sort of comprehensive, national legislation is critical to the future of the industry.

"We need a national recycling policy," says Edgar Miller, director of policy and programs for the National Recycling Coalition (NRC, Washington, DC). "We have to put recycling back on Congress' radar screen."

At its May 1994 meeting, the NRC board voted 24 to two (with two members abstaining) to implement a full-scale national recycling act advocacy program. "The vote was a strong statement which reinforces the NRC's efforts to develop a more significant advocacy role," Miller says. "There is a desire on the behalf of the board to help NRC realize its potential as a national voice on recycling.

"We think there are certain issues that can only be addressed at the national level," he continued. According to Miller, NRC believes a national policy is needed to:

- Reduce waste, conserve natural resources, and increase the utilization of recovered materials;
- Ensure the development of nationally uniform standards, definitions, and public policies;
- Define and establish national waste reduction and recycling goals;
- Expand the recycling infrastructure;
- Go beyond current levels of recycling, where market forces are not working; and
- Define the roles and responsibilities of major stakeholders.

"We're going to be leaders in this issue," says Harry Benson, chair of NRC's Policy Research Committee and manager of market development for Wellman, Inc. (Shrewsbury, NJ). "We've always had bits and pieces of this program. We will bring forward a comprehensive, broad-based platform and . . . get it out and on the ears of the policy-makers in D.C."

Though still in its preliminary stages, the proposed advocacy message will address demand- and supply-side policies; policies to make recycling systems more economical; labeling policies; and education strategies.

However, how those objectives will translate into recycling legislation remains to be seen. "The bill may take the form of national recycling rate legislation, or they might want to have minimum-content standards," says Marty Forman, [then-]president of Poly-Anna Plastics (Milwaukee) and NRC board member.

Or, "it could mean a full-time lobbyist on Capitol Hill. We're still in the process of clarifying that," explains Cynthia Conklin, principal of Resource Recycling Systems (Ann Arbor, MI) and NRC board member, adding that a myriad of options are being considered.

However, coming to a consensus on which agenda to push presents a challenge to NRC. "One of the keys is our ability to come to some agreement on a demand-side legislative policy," Conklin says. "The bottom line is that we need to actively communicate the priorities of recycling to those at the national levels [but] there are certain issues that a coalition will not come to quickly," she adds.

"[This initiative] is something whose time has come," Benson says.

The Federal Example

Over the past two years, however, the most notable federal endeavor in regard to recycling was President Clinton's executive order, which instituted guidelines for federal agency acquisition of products comprised of recovered materials.

Federal environmental executive Fran McPoland, appointed in June 1994 to implement the executive order, says that she hopes that the "federal government will drive markets for recyclables by leading by example." In fact, states such as California and New York have begun to promote similar purchasing directives.

Some industry officials, however, see the impact of the executive order as negligible, and are calling for even less government intervention on both the state and national levels. The executive order "has almost incidental value as far as reducing waste is concerned," says Terry Bedell, environmental packaging manager for Clorox Co. (Pleasanton, CA). "The more government tries to control market conditions, generally the worse it gets."

Still, some environmentalists and industry officials would prefer that the federal government mimic the packaging initiatives of

Germany and Canada, thus implementing even more comprehensive regulations at the national level.

Foreign examples, such as the aggressive German Green Dot Program, [see Chapter 8] have spurred discussion of similar national recycling legislation in the U.S. But American legislators have generally preferred to take a "wait and see" stance.

Both Sen. Max Baucus (D-MT) and Rep. Al Swift (D-WA) have commented on the Green Dot's sweeping agenda and how such a program could apply to the U.S. "While European recycling programs cannot be adopted directly in America, there are some ideas that I believe will prove useful for us," Baucus said at a 1993 meeting of the U.S. Conference of Mayors and National Association of Counties.

"We should go to a Green Dot-like system very, very carefully," Swift added at a recent recycling debate sponsored by *The Atlantic Monthly.*

A national recycling bill, similar to the Green Dot in terms of its emphasis on manufacturers' responsibility, had been proposed by Baucus in the 102nd Congress, but failed. Still, Baucus resurrected the issue in 1993. "The cornerstone of my strategy rests on the principle that I call 'manufacturers' responsibility' for the life-cycle of a product," Baucus said. Other options included in his proposed initiative were:

- minimum recycled-content standards;
- a waste utilization tax; and
- standardization of federal and state procurement regulations.

Despite the Senator's efforts, however, the "Baucus bill" never came to fruition. "The cities which had pushed for some recycling legislation last [102nd] Congress have not this [103rd] Congress," explains a spokeswoman for Baucus. "The cities have other environmental issues that they are more concerned about, . . . but [Baucus] is very interested in recycling . . . and he'll be back."

Bottle-Necking Legislation

Part of the problem with getting a national bill passed resides in the divergent viewpoints within the recycling industry itself as to which means should be enforced to achieve the end.

One example of this kind of impasse can be seen in the debate over passage of a national bottle bill. Early in 1994, the proposal, which would impose a 10-cent deposit on bottles and cans in states that do not recycle at least 70% of such containers, had gained attention from major, bipartisan backers in Congress. Senators Mark Hatfield (R-OR) and Jim Jeffords (R-VT), and Representatives Edward Markey (D-MA) and Fred Upton (R-MI) had agreed to skip the committees and tack on a bottle bill as an amendment to a larger interstate waste restriction bill. By year end, however, support for a bottle bill amendment had faded considerably, especially in the Senate, due to fears the amendment would hamper efforts to pass the politically hot waste transportation bill.

Baucus, though an outspoken proponent of national recycling legislation, voted against an amendment to the Resource Conservation and Recovery Act (RCRA) that would have created a national system for returning glass beverage containers.

Other recycling advocates, however, view such legislation as a way, when combined with curbside programs, to divert more waste from landfills. "We strongly support a national bottle bill because [bottle bills] have proven to be the most effective recycling legislation in this country," says Gene Karpinski of the U.S. Public Interest Research Group (Washington, D.C.). "When combined with a curbside program, you get the best of both worlds," he adds.

Once again, however, the bill is not expected to go anywhere in either the Senate or the House.

Predictions

Opinion among most industry and congressional sources is that RCRA will not resurface until 1996, though no one has ruled out the possibility of an earlier emergence.

"If the other environmental acts like Superfund . . . stumble in the reauthorization efforts in the next couple of years, there may be a shift to RCRA because of the [1996] presidential election," explains Jonathan Greenberg, director of environmental policy for Browning-Ferris Industries, Inc. (Houston, TX). "A lot depends on the economy and local government, and whether they feel that they need an extra push," he adds.

In any case, when RCRA does resurface, many in the industry want to be ready with something to bring to Congress' table. "We should have a targeted message for RCRA reauthorization if that ever becomes a priority," Conklin says. "We know what we have to do: We have to get our message out there."

3 WHAT DOES RECYCLING REALLY COST?

As the enthusiasm for recycling spread throughout the U.S. in the late 1980s and early 1990s, concern about the cost of recycling also became widespread. But the movement had grown so quickly that data were sparse. In 1992 and 1993, the National Solid Wastes Management Association funded two studies: one on the cost of processing recyclables and the other on the cost of collecting recyclables. The article summaries of these studies are reprinted here.

THE REAL COST OF PROCESSING RECYCLABLES

By *Chaz Miller*, manager of recycling programs for the Environmental Industry Associations, Washington, D.C.

How much does it actually cost, on the average, for a typical materials recovery facility (MRF) to process one ton of recyclables?

(A) About $5
(B) About $25
(C) About $50
(D) About $100

The correct answer is C, according to a 1992 study from the National Solid Wastes Management Association (NSWMA, Washington, DC). To determine the real costs of MRFs, NSWMA sponsored "Processing Costs for Residential Recyclables at Materials Recovery Facilities," a survey that took a thorough look at 10 facilities and what they now pay to process recyclables.

MRF Technology Evolved

As residential recycling programs increase in size and complexity, more and more communities are opting to collect recyclables by commingling rigid containers and then "demingling" them at a MRF. While commingled collection began in the mid-'70s, it blossomed in the late '80s. MRFs, originally called intermediate processing centers, evolved from crude operations that handled old newspapers (ONP), metals, and glass, to multifaceted processors of several grades of paper, glass, metals, and plastics. While knowledge of the technologies involved in MRFs has steadily increased, the economics of MRFs has remained a mystery.

Some attempts at analyzing MRF economics have already been made. Virtually every consultant's study analyzing local recycling options estimates the cost of operating a MRF in a particular location. In addition, engineering estimates or the results of surveys can be used to establish the cost of MRF operations. However, estimates and surveys do not give "real world" data. Surveys could be completed inconsistently, and engineering estimates are just that—estimates.

As a result, NSWMA's Waste Recyclers Council (WRC), which is composed of businesses engaged in the collection and processing of recyclables, decided to fund a study to determine the cost of processing materials. MRFs, after all, can represent a significant capital investment. It is not unusual to invest more than $10 million in capital costs in a MRF, and the trend is for larger, more expensive facilities. The Roy F. Weston consulting firm was selected to conduct the study; Dan Briller and Abbie Page (of MacMillan Consultants) led the project team.

WRC wanted to focus on MRFs in the 100–300 tons per day (tpd) size. These MRFs, it was thought, represent the most common size of MRFs operating in the U.S. In its *1992–93 Materials Recovery and Recycling Yearbook*, Government Advisory Associates (New York, NY) lists 131 tpd as the average size for MRFs. WRC also decided to focus on privately-operated facilities, largely because it was believed that the data from private facilities would be more reliable and more complete, although the consultants would be asking for highly proprietary business information concerning operating costs. To ensure confidentiality of individual data, facilities were guaranteed that all data sheets and computer disks would be destroyed at the end of the study, and that no references would be made in the report to individual facilities. Even with those assurances, several MRF operators declined the opportunity to participate in the study because of a fear that confidential data would be inadvertently revealed. Ten facilities, with an average throughput of 162 tpd, agreed to participate. These facilities are located in the Northeast, South, West, and Midwest.

MRF Methodology

Before selecting participants, WRC's MRF committee devised a cost allocation format. While it is relatively easy to determine the average processing cost for a ton of recyclables, it is not as easy to determine the average processing cost for a particular recyclable. For example, both equipment and labor costs can be assigned to more than one material. Another example is negative sorting, which is a good way to lower labor costs by leaving materials on the conveyor, but it poses allocation problems when counting the costs of processing.

An equitable system was devised, however, to assign processing costs to individual materials in the MRF process, from space on the tipping floor to an allocation for residue. Where possible, costs were assigned to individual recyclables (e.g., a baler dedicated to ONP can be directly assigned to ONP, and a baler used for both ONP and plastics can be allocated, based on the time spent on each material). When costs could not be directly allocated, they were assessed, with one exception, on a weight basis. Building costs—land or rent—were allocated on a volume basis. In that situation, the consultants first estimated how much building space was assigned to processing paper and how much to processing commingled containers. Then the commingled container section (usually 50% of the building space or less) was given the volume allocation shown in Table 3.1.

All 10 facilities were visited to conduct the study. Prior to the visit, they received a detailed questionnaire. The consultants used the site visit as an opportunity to verify information and fill in missing data.

Table 3.1 Volume Allocation of Recyclables

Aluminum cans: 12%	Steel cans: 16%
Clear glass: 7%	Brown glass: 4%
Green glass: 3%	Mixed glass: 5%
High-density polyethylene (HDPE): 29%	
Polyethylene terephthalate (PET): 21%	
Residue: 3%	

As expected, facilities varied in their ability to fill out the questionnaire. Some were able to fill out all questions without any problem, while others required follow-up phone calls to fill in data and verify items. Eventually, all 10 questionnaires were consistently completed. As a result, the data cover all processing costs at the participating MRFs.

These data, however, specifically do not include revenues from the sale of recyclables or profit/loss calculations. This is because revenues and profit/loss calculations affect the bottom line at a MRF, but they do not affect processing costs.

Two items created accounting problems. Land costs could not be assigned for three facilities because they are located on land that is already used for solid waste management activities, such as transfer stations or landfills. Opportunity costs for the use of this land would be highly speculative, at best. However, a MRF processing 160 tpd on a five-acre site with a land cost of $250,000 would see an additional processing cost of 20 cents per ton—assuming 30-year, straight-line depreciation.

Finally, in order to avoid hidden subsidies, residue disposal costs based on local tipping fees were assessed for three publicly-owned facilities with "free" disposal at nearby public landfills.

Costs Outweigh Revenues

According to WRC's study, the average cost of processing a ton of recyclable materials at a MRF, before sales revenues are considered, is $50.30. The range of processing costs at these 10 MRFs went from a low of $28.11 per ton to a high of $72.06. Because the average revenue from a ton of commingled recyclables that have been processed is $25–30, it is clear that tipping fees or contractual provisions with supplying communities are necessary to allow MRF operators to cover their costs and make a reasonable profit.

After determining the cost to process a ton of commingled recyclables, the study then determined the cost differences of processing each type of recyclable. Paper is less costly to process than commingled materials. The average cost of processing a ton of paper was $33.5 with a range of $20.43 to $55.93 per ton. Commingled materials, by contrast, cost an average of $83.36 per ton to process with a range of $40.76 to $146.29.

Table 3.2 Processing Costs by Material (Costs in dollars per ton)

Material	Processing Cost	
	Average	Low–High
Newspaper	33.59	19.94–55.33
Corrugated Material	42.99	20.29–56.26
Mixed Paper	36.76	16.82–65.59
Aluminum Cans	143.41	72.88–362.59
Steel Cans	67.53	30.22–125.64
Clear Glass	72.76	37.17–105.62
Brown Glass	111.52	69.70–148.92
Green Glass	87.53	57.56–134.21
Mixed Glass	50.02	28.51–76.24
PET Plastic	183.84	64.43–295.79
HDPE Plastic	187.95	121.58–256.15

Source: National Solid Wastes Management Association, "Processing Costs for Residential Recyclables at Materials Recovery Facilities," 1992.

Of course, the cost of processing each type of material within a commingled stream varies greatly. For example, it costs an average of $72.76 to process one ton of clear glass, with a range of $37.17 to $105.62 per ton. A ton of PET costs an average of $183.84, with a range of $64.43 to $295.79. Newspaper costs an average of $33.59 per ton to process, with a range of $19.94 to $55.33 (see Table 3.2 for material processing costs). Generally, potential revenues for these commodities are below the average processing costs, although facilities at the low end of the cost scale may be able to profitably process these materials.

The cost of processing paper is relatively low, both in absolute terms and compared to processing costs for other recyclables. Paper represents 50–75% of the throughput at a MRF. Its volume and its relatively limited reliance on labor make paper the most "productive" material at a MRF, with an average employee productivity of 7.42 tons of paper processed per employee per day (see Table 3.3 for employee productivity). As a result, increasing the amount of paper recycled will lead to lower overall processing costs.

Not all types of waste paper are cost-effective to process at this point, however. Mixed paper, which represents the largest amount of unrecycled paper, is a prime example. With the high costs involved in upgrading mixed paper to revenue-rich paper grades, and the negative value of mixed paper in most markets, it doesn't make sense to add

Table 3.3 MRF Employee Productivity (Productivity in tons per employee, per day)

Material	Average	Low–High
Paper	7.2	4.8–13.0
Metals	5.96	1.8–17.5
Glass	4.21	2.0–10.0
Plastics	1.57	1.0–2.5
Total	5.04	2.65–8.31

Source: National Solid Wastes Management Association, "Processing Costs for Residential Recyclables at Materials Recovery Facilities," 1992.

mixed paper to a recycling program without strong local markets for the end product.

Factors Affecting MRF Costs

A number of factors affect processing costs at a MRF. While larger facilities that operate at, or near, design capacity and that make efficient use of labor tend to have lower costs, they were not always the lowest-cost facilities.

Labor is the highest cost, accounting for an average of 33.4% of MRFs' overall processing cost components (see Table 3.4). Labor costs in the 10 MRFs studied ranged from 27.1% to 43.3% of total

Table 3.4 Primary MRF Cost Components (Expressed in percentage of total cost)

Cost	Average	Low–High
Labor	33.4	27.1–43.3
Building rental/amortization	16.7	7.5–34.4
Equipment amortization	13.5	6.0–25.2
General administration	13.0	2.1–28.2
Insurance, workmen's comp., misc.	9.4	
Residue hauling/disposal	7.7	1.5–16.6
Maintenance/repairs	6.1	2.0–14.8

Source: National Solid Wastes Management Association, "Processing Costs for Residential Recyclables at Materials Recovery Facilities," 1992.

processing costs. As a result, MRF operators are constantly looking for ways to substitute equipment for labor. Usually this is successful, but all materials have thresholds below which the use of equipment to process them is not cost-effective.

Eddy current separators, for example, can lead to lower processing costs for aluminum used beverage cans, but a facility needs a minimum amount of aluminum to process in order to justify the capital costs of such a machine. MRF automation has also gained popularity and increased processing efficiency in such areas as plastic sorting. However, materials specifications and technologies could always change, potentially rendering the automated systems less efficient and the investment made in new equipment a less productive use of capital.

Negative sorting, or processing a material by not touching it but letting it move on the conveyor to fall into its own hopper, leads to lower processing costs for that material by eliminating labor.

Clear glass, for instance, was negatively sorted at five of the facilities in this study. Clear and mixed glass are generally less costly to process than green or brown glass. This is because they are more abundant in the waste stream and are often negatively sorted or removed mechanically by a screening machine. This also implies that materials recovered in lesser quantities via positive sorting may be more costly to process. Negative sorting, however, can only be used for a limited number of materials.

Building rental or amortization is the next greatest cost, averaging 16.7% of the MRFs' total processing costs. The 10 MRFs studied reported a range of 7.5% to 34.4% of total processing costs that go toward rent or amortization. Locating a facility on land currently used for solid waste management activities, therefore, may lower land costs and avoid NIMBY battles over siting MRFs.

Equipment amortization costs are the next most expensive part of MRFs, with a range of 6% to 25.2%, and an average of 13.5% of total processing costs. Sales and general administration costs ranged from 2.1% to 28.2% of the processing costs for these MRFs, or an average of 13%. Residue hauling and disposal can be another problem area, with a range of 1.5% to 16.6% of processing costs at the 10 MRFs, and an average of 7.7%.

Residue costs can be kept down, for example, by finding markets for the mixed glass portion of the residue. While these markets—usually aggregate markets—will have little per-ton value, they can provide a small cash flow and avoid the need to pay a landfill tipping fee.

In addition, improved collection and processing will lead to lower amounts of residue and lower overall processing costs.

Betting on the Market

This MRF cost study proves that it does cost money to process residential recyclables. Based on these costs, and current revenues from end markets, few materials can be profitably processed. This should come as no surprise to anyone involved in solid waste management. Tipping fees—or their counterpart, processing contracts—are necessary at most, if not all, MRFs to cover processing costs.

In addition, markets must be developed so that materials can be economically recycled. Requiring the collection of recyclables without equal requirements for use of the collected materials leads to weak markets and unprofitable recycling programs. When manufacturing industries accept responsibility for providing markets for their products, tipping fees at MRFs may not be needed.

Processing does not take place in a vacuum, and this study is really a snapshot of what is currently going on at MRFs across the country. New technologies and ideas could increase the cost efficiencies, and therefore decrease processing costs at some MRFs. But if markets continue to remain weak, and if specifications continue to tighten as a result, then processing costs could increase.

****

THE COST OF RECYCLING AT THE CURB

How Much Does It Cost to Collect One Ton of Recyclables at the Curbside?

(A) $90 (B) $115 (C) $120 (D) $150

By *Chaz Miller*, manager of recycling programs for the Environmental Industry Associations, Washington, D.C.

The correct answer, according to most surveys, is all of the above. The answer, derived from a 1993 study from the National Solid Wastes Management Association (NSWMA, Washington, D.C.), is that the cost for a truck and a crew to collect commingled residential recyclables five days a week is between $104,000 and $148,000 per year. Route cost does not include processing cost, revenue from the sale of recyclables, or the cost of containers.

Per-ton costs vary, depending on crew size, truck capacity, set-out rate, distance between stops, and other factors. On a "typical" suburban route, per-ton cost will probably be between $115 and $120 per ton.

To gain better insight into the costs of collecting recyclables, NSWMA's Waste Recyclers Council (WRC) sponsored "Collection Costs for Residential Commingled Recyclables," a look at the costs involved in collecting recyclables. Unlike the WRC's previous study, "Processing Costs for Residential Recyclables at Materials Recovery Facilities," this is not a full-scale look at operational data. The materials recovery facility (MRF) processing universe is small and relatively homogeneous. In contrast, the commingled collection universe is large and diverse, with many variables.

As a result, the WRC decided to determine the yearly cost of operating a recycling collection route and then apply those costs to a "typical" suburban route, testing the effect of variations in crew size, truck size, and set-out rates. The WRC hired Killam Associates (Millburn, NJ) and Ecodata, Inc. (Westport, CT) for the project. Ed Jablonowski (Killam) and Barbara Stevens (Ecodata) were the lead consultants.

WRC members wanted to understand all the costs involved in collecting recyclables. They wanted a complete list of recycling collection costs. They also wanted to know which collection variables were the most important. This led to the decision to use the "full-cost accounting" methodology to allocate costs. All the operating and capital costs applicable to recycling—including collection and processing equipment, labor, buildings, land, administration, and overhead—would be included. Another reason for using full-cost accounting is that several states require it for all solid waste management systems.

Yearly Crew Costs

The first step was to determine the yearly operating and capital cost on a collection route. Route costs are the basis of collection costs and can be extrapolated based on a community's size. Using the Killam-Ecodata database and the experience of WRC members, crew and vehicle costs, operation and maintenance, building, administration, and overhead costs were developed. (Surveys of operating programs were rejected because of the immense diversity in programs and the reality that surveys are often answered inconsistently, with many costs ignored. Failure to include equipment depreciation, capital cost, and administrative and overhead costs lead to inaccurate results.)

Costs were established for both one-person and two-person crews using either a 23-cubic-yard truck or a 31-cubic-yard, dual-side-loading truck. Yearly costs for a recycling route ranged from $104,000 to $148,000, depending on crew size and truck capacity (see Tables 3.5 and 3.6 for a complete listing of crew costs). Labor is the biggest cost, followed by vehicle operation, and maintenance and vehicle depreciation. Together, labor and vehicle costs are about 80% of total costs (see Table 3.7).

The Typical Suburban Route

These costs were then applied to a typical suburban route. Variables, including set-out rate, and crew and vehicle size, were tested.

Developing a typical route has obvious pitfalls. Computer models are limited to the assumptions used in fashioning the model. However, they can be very useful in testing variables and learning which are the most important.

But, what is a typical route? America is a large country with thousands of communities of different sizes and geography. The WRC typical route is based on WRC member experience and the Killam/Ecodata database. It is no more or less than a model designed to test variables. Readers should take into account the unique circumstances of their own routes.

In the WRC's typical route, newspaper is set out in a bundle. Commingled glass bottles, aluminum and steel cans, polyethylene terephthalate soft drink bottles, and high-density polyethylene milk cartons are placed in a bin. The start-up to on-route time is 30 minutes, with 60 minutes for getting to and from the processing site. Residential units are predominately single-family. Distance between households is 150 feet.

One-person crews collect from both sides of the street 50% of the time, while two-person crews collect from both sides of the street 90% of the time. (The study does not endorse two-sided collection. Some companies will not allow two-sided collection due to safety concerns.) The standard work week is five days, 8.5 hours per day, with a base hourly wage of eight hours plus a half hour of overtime.

Table 3.5. Crew Costs

Cost Element	One Person $/Year	Two Person $/Year
LABOR & FRINGES		
Foreperson (10%)	$ 3,000	$ 3,000
Driver	25,000	25,000
Helper/Driver	0	18,000
Backup[a]	2,692	4,423
Overtime[b]	2,625	4,313
TOTAL LABOR	33,317	54,736
FICA (6.2% on 57.5)	2,066	3,394
Medicare (1.45% of salary)	483	794
Workmen's Comp (8.5%)	2,832	4,653
Retirement (0–11%, Med. 5.5%)	1,832	3,664
Health/Life/UI/DI Ins. (5–20%, Med. 12.5%)	4,164	8,328
Clothing Allowance	220	440
TOTAL FRINGES	11,597	21,273
VEHICLE OPERATION AND MAINTENANCE		
Mechanic's Wages & Fringes[a,g]	7,381	7,381
Parts	5,000	5,000
Fuels & Fluids	9,361	9,361
Insurance	5,000	5,000
Licenses & Taxes	1,000	1,000
Backup Vehicles[c]	2,774	2,774
Mobile Equipment[d]	1,467	1,467
TOTAL OPERATION AND MAINTENANCE	31,983	31,983
BUILDING AND UTILITIES[e]	3,814	3,814
OTHER EXPENSES		
Employee Training	1,000	2,000
Public Education	2,033	2,916
Property Damage	1,200	1,200
Miscellaneous[f]	36	512
TOTAL BUILDING/UTILITIES/OTHER	8,083	10,442
VEHICLE DEPRECIATION		
23-Yard capacity	7,857	7,857
31-Yard capacity	13,986	13,986

Table 3.5 Crew Costs (Continued)

Cost Element	One Person $/Year	Two Person $/Year
SUBTOTAL		
23-Yard capacity	92,838	126,291
31-Yard capacity	98,966	132,419
ADMINISTRATIVE 12%[g]		
23-Yard capacity	11,140	15,155
31-Yard capacity	11,876	15,891
TOTAL ROUTE COST		
23-Yard capacity	103,978	141,446
31-Yard capacity	110,842	148,310

[a]Backup labor at 25/260 of regular, for absences of all types, including vacations.

[b]Overtime at 1.5 times regular pay, for one-half hour per day.

[c]One per 10 first-line vehicles.

[d]$15,000 vehicle operating expense, for 30 vehicles; salary and fringes for driver at $29,000.

[e]See Table 3.6 for included items.

[f]Reflects functional inefficiencies of two-person crews.

[g]Administrative overhead, including customer services, management, personnel, and all other overhead services. Overhead costs are 8.7% for nondepartmental functions, plus departmental overhead (dispatch, routing, customer services, etc.) at 3.3%.

Table 3.6 Building and Utilities

Cost Element	Amount/Year	Amount/Crew
Facility Rental[a]	$24,000	$2,400
Utilities		
Oil (heat)	3,000	300
Electricity	3,000	300
Telephone	3,600	360
Office Supplies	2,400	240
Furniture & Computers		
$15,000 depreciated over 7 years	2,143	214
TOTAL	38,143	3,814

Footnotes:

[a]One-half acre, with 2,000-square-foot office, 4,800-square-foot maintenance/utility building and parking.

Table 3.7 Cost Distribution

Cost Element	1-Person Crew	2-Person Crew
Labor & Fringes	38–46%	47–56%
Vehicle O&M	30%	22%
Building, Utilities	4%	3%
Other Expenses	4%	5%
Vehicle Depreciation	7–13%	5–10%
Administration	12%	12%

(This pay formula, along with assumptions about the percentage of time needed to collect materials from both sides of the street, strikes a balance between operating conditions throughout the country.)

Tonnage collected in the various scenarios ranged from less than three tons a day to more than five tons a day, with an assumption of an average set-out of 11.45 pounds of recyclables. Daily route size varies based on set-out rate, with higher set-out rates resulting in smaller routes.

Per-Ton Costs on the Typical Route

Crew productivity and the percentage of households setting out recyclables are the keys to per-ton costs. Productivity was measured by a number of factors including available on-route time; travel time between set-outs; loading time (relying heavily on Killam-Ecodata's database); low (25%), medium (50%), and high (75%) set-out rates; and collection truck capacity. The 50% set-out rate is the closest to normal operating conditions.

Clearly, picking up more material from more stops per day and spending less time driving from stop to stop makes a big difference in per-ton costs. For instance, the two-person crew cost ranges from $148.77 per ton on the 31-cubic-yard truck with a 25% set-out rate to $104.30 per ton with a 75% set-out rate. At 50% set-out, the cost is $114.12 per ton (see Table 3.8).

Surprisingly, a two-person crew can be more productive and less expensive than a one-person crew. This is because the model used a dual-side-loader truck with the crew collecting from both sides of the street 90% of the time. If the crew collects from both sides of the street only 50% of the time, per-ton costs increase. If, for safety concerns, a company will not allow two-sided collection, costs will increase further.

Table 3.8. Route Productivity and Costs for Dual-Side-Loading Trucks

A = 23-cubic-yard truck B = 31-cubic-yard truck

	Set-Out Rate 25%		50%		75%	
	A	B	A	B	A	B
ONE-PERSON CREW						
Cost/HH/Mo.	$ 0.95	$ 0.98	$ 1.50	$ 1.50	$ 2.01	$ 2.08
Cost/Ton	153.76	158.60	120.71	120.98	108.19	111.65
Cost/Cu. Yd.	23.44	24.18	18.40	18.44	16.50	17.02
Cost/Stop	0.88	0.91	0.69	0.69	0.62	0.64
TWO-PERSON CREW						
Cost/HH/Mo.	$ 0.91	$ 0.92	$ 1.44	$1.42	$1.91	$1.94
Cost/Ton	146.59	148.77	116.10	114.12	102.90	104.30
Cost/Cu.Yd.	22.35	22.68	17.70	17.40	15.69	15.90
Cost/Stop	0.84	0.85	0.67	0.65	0.59	0.60

NOTE: These costs are based on assumptions used in creating a ‘typical’ suburban route.

Other factors affecting productivity include collection frequency, the number of set-out receptacles per stop, the number of material separations at the truck, routing, inadequate supervision, and housing density.

Per-Material Costs

The WRC also looked at per-material collection costs. These costs should not be interpreted as hard-and-fast costs, but as indicators of relative ranking in cost. Glass and paper were the least expensive to collect, largely because of their weight. Steel, aluminum, and plastic were the most expensive to collect (see Table 3.9). However, on a marginal cost basis, the cost to add plastic to a collection program is $224 per ton in the small truck and $107 in the larger truck (assuming a 50% set-out rate).

What Does It Mean?

Recycling costs money. This should come as no surprise. While the best things in life may be free, recycling (like all forms of solid

Table 3.9 Per Ton Costs of Collecting Individual Materials

Material	Set-out Rate 25%	50%	75%
Newspaper	$ 93	$ 72	$ 65
Glass	77	60	54
Steel	309	240	217
Aluminum	748	581	526
Plastic	1,401	1,089	987

NOTE: These costs are based on assumptions used in creating a 'typical' suburban route.

waste management) isn't. According to prices quoted in *Recycling Times* [when this was first written in 1993], a ton of commingled recyclables is worth less than $40.

As recycling collection operators become more experienced, collection prices will rise to offset underestimated costs and overestimated revenues. Seattle offers an interesting example of this. Contract prices for collecting and processing recyclables increased from an average base price of $48 per ton in 1988 to an average base price of $81 in 1993. In 1988, one contractor had a revenue risk-sharing provision with the city—now both contractors do. Inflation is part of the reason for the increase in the base price. A bigger reason is that the contractors learned enough about recycling to bid more intelligently.

Yet, in spite of increased cost, recycling is still popular in Seattle. According to NSWMA polls, most Americans want to recycle and are willing to pay for it. As recycling programs become more efficient, as more households set out more materials, per-ton collection costs will decrease. Recycling program operators, however, must cover their costs and make a profit. Better markets for recyclables and increases in collection productivity are a must. Current market development efforts are beginning to expand some markets. Given a healthy economy and expanded markets for recyclables, recycling will hold its own.

CODA

Per-Household and Avoided Costs: Magic Money or Real Savings?

Solid waste management costs money. Recycling has collection and processing costs. Disposal, either through landfilling or incineration, has collection, transfer station, and tipping-fee costs. Recycling and incineration, however, can offset some of their costs with revenues from material or energy sales. According to a 1988 study, when prices for recyclables—especially newspaper and aluminum cans—were higher than they are today, revenues covered less than half the cost of collecting and processing recyclables. While a more recent study showed recycling to be less expensive than solid waste collection and disposal in four of the biggest cities in the state of Washington, this was caused in part by relatively high tipping fees in the Pacific Northwest. According to published figures, tipping fees are even higher in the Northeast and lower in most of the rest of the U.S.

Recycling advocates often promote noncash factors. These include avoided collection and disposal costs and lower per-household collection costs for recycling than for solid waste. But are these avoided costs real, or are they just magic money? And are lower per-household collection costs a sign of strength or weakness?

Let's look at per-on-route-household collection cost first. Several studies have shown a lower per-household collection cost for recycling than for solid waste. For instance, one New Jersey study showed per-household costs for curbside recycling to be only one-seventh as much as per-household costs for solid waste collection and disposal. And why not? On an average route, most households set out solid waste, while perhaps half set out recyclables (recycling programs usually have high monthly participation rates, with lower weekly participation rates, while household participation in solid waste collection is high on a weekly basis). Recycling trucks spend more time driving and collect less material than solid waste trucks.

Cost per on-route household is highly sensitive to participation rates. When recycling routes start collecting more materials from more stops, the size of the routes will diminish and the cost per on-route household will increase. The WRC study showed increases in per-household costs and decreases in per-ton costs as the number of set-outs increased (see Table 3.8). Per-stop costs, likc pcr-ton costs, diminish as the number of stops and the amount of recyclables collected increase.

Avoided disposal costs are often cited as a major economic benefit for recycling. According to this theory, recycling programs should be credited with the disposal cost of each ton of material collected for recycling. If, say, tipping fees (including transfer-station costs) are $50 per ton, that amount should be deducted from the cost of recycling. After all, didn't the recycling program save the tipping-fee cost?

Yes and no. Avoided costs raise a number of problems. The first is the tendency to credit a collection program with an avoided cost for each ton of recyclables it collects. Yet, paper drives by charitable groups, drop-off centers, and aluminum can buyback programs are all examples of preexisting recycling collections that get displaced by, or compete with, curbside collection programs. Certainly, more recyclables are collected as a result of curbside programs. Crediting a curbside program with avoided disposal costs for each ton collected denies the preexistence of other recycling collection systems.

However, avoided disposal costs have an even more serious flaw. Because of the recycling program, money was not spent on the tipping fee. But it was spent on the recycling program. Suppose collection and disposal of solid waste costs $100 per ton (including a $50 tipping fee) and the recycling program costs $100 per ton (including collection and processing costs, less revenues). In this case, while the recycling program avoided the $50 per ton transfer station/disposal tipping fee, it spent the money on recycling. The bottom line is that both options cost $100 per ton. The total amount spent on solid waste management will be the same with or without recycling. What appears to be a saving is merely a shift in cost from one budget column (tipping fees) to another (recycling). The same will hold true if recycling is more expensive after revenues are deducted. Only if the recycling program is less expensive, are true savings realized.

Occasionally, curbside recycling gets credited with avoided collection costs. In this case, the theory is that recycling lowers the cost of collecting solid waste. In some cases, this is true: Twice-a-week solid waste collection may be reduced to once-a-week collection, or curbside recycling may have eliminated enough solid waste to allow for rerouting of trucks and elimination of whole routes.

However, if a community already has efficient solid waste collection, curbside recycling is unlikely to result in real collection-cost savings. Many of the cited instances of savings are examples of more efficient solid waste collection that may have had nothing to do with the curbside program or that may have been prompted by a desire to improve overall efficiency of solid waste collection in order to add collection of recyclables to the system.

4 MARKETS

RECYCLING COMING OF AGE: RECYCLING MARKETS, 1990–1994

By *Lisa Rabasca*, editor of *Recycling Times*, Washington, D.C.

In five years, recycling has gone from a nascent business to a maturing industry.

What began in the late 1980s as a business trying to get off the ground, is now recognized by many cities and states as a means of economic development and job creation.

But recycling's coming of age was not without growing pains. Many recyclers had to hang on while markets were poor and spotty. Gluts of plastic, wastepaper, aluminum, and green glass often made it difficult for recyclers to turn a profit. Until early 1994, prices for most commodities were significantly low, and in some case, these low prices forced recyclers and processors to close their doors or at least curtail their operations.

During summer 1994, markets improved substantially as prices for wastepaper, plastics, and aluminum began to soar. By the end of 1994, it was not uncommon for mills to pay more than $100 per ton for old newspaper (ONP). This was a dramatic change from 1992, when processors often had to pay mills to take their ONP.

As 1995 approaches, mills and plastics plants find themselves scrambling for enough feedstock, paying top dollar for material that only two years ago they often could receive for free. Today, many in the industry wonder if enough material can be collected to keep domestic mills and plants running at capacity.

Modest Beginnings

Despite recycling's upturn, the industry had modest beginnings. Initially, most recycling focused on ONP—a commodity which is the easiest to collect and recycle.

As a result, markets for ONP suffered from severe oversupply. Domestic capacity for recycling ONP was yet to come on-line and export markets were weak in 1989. Wastepaper exports to South Korea fell in 1989 because of internal political changes to the country. At the time, South Korea had been the biggest overseas buyer of the U.S.'s ONP. In July 1989, the price of ONP exported from the port of Newark, NJ, dropped 30% to $12 per ton from $17 per ton in June. The previous month, the price had been in the $20-per-ton range.

On the West Coast, falling export prices for ONP forced one private recycler in California to decrease its waste paper collection by 20%. "In my opinion, if the price of paper is below $40 per ton, recyclers cannot make money handling it," Steve Moore, president of Pacific Rim Recycling (Benicia, CA) said in August 1989.

Southeast Paper Manufacturing Co. (Dublin, GA) cut back on buying ONP in August 1989, despite its plans to nearly double its capacity for making recycled newsprint.

Weak end-markets for ONP often forced processors to cut their operations. Philadelphia's main processor of recyclables had to shut down for three months in 1989 because of an oversupply of ONP and plastics. Basically, the processor had collected too much material too soon from curbside programs.

Chicago started its recycling program in 1989 without ONP collection, while Washington, D.C., collected ONP but could not find a market for its material.

Looking Toward the Future in 1989

As more municipal and commercial recycling programs came online, industry experts began to make predictions about the future of recycling. Some projections were on target; others were not.

Bleak markets in June 1989 caused one Baltimore wastepaper broker to predict that ONP prices would remain weak for at least three to four years. The broker suggested that municipalities stop collecting wastepaper, especially ONP, because of the glut on the market.

Another industry expert agreed that ONP prices would gain strength in three to four years, but he urged municipalities to continue collecting wastepaper.

Not all predictions came to pass. One expert said that mixed wastepaper would be "the dog" of the recycling industry because there are no end-uses for it. Today, mills are concerned that there will not be enough mixed wastepaper collected to feed capacity in the year 2000.

Volatile Markets

Going into 1990, prices for recyclables—particularly wastepaper—were low. Too many office wastepaper recycling programs coming on-line simultaneously caused a glut of material on the market. Asian importers announced price changes in March, and pulp prices were so low that mills preferred to buy pulp rather than pulp substitutes such as computer printout (CPO) and white ledger.

To make matters worse, overseas shippers had to contend with an $11 per ton "bunker charge" to cover fuel costs stemming from the Persian Gulf fuel crisis. In 1990, according to *Recycling Times* data, mills paid $11 to $22 per ton for ONP and processors paid –$3 to $7 per ton.

Glass prices also came crashing down. Glass manufacturers had been buying glass at an artificially high price to encourage recycling. Suddenly, end-users were lowering their base price, and processors paid about $11 to $25 per ton for clear glass in 1990.

Prices for polyethylene terephthalate (PET) increased in the spring but, by summer, prices fell because worldwide polyester fiber prices were down. PET processor prices ranged from 6.6 cents to 8.5 cents per pound in 1990.

Aluminum saw the largest fluctuation, with prices rising in spring but then declining by summer. The toll price for aluminum used beverage cans (UBCs) ranged from 38 cents to 54 cents per pound in 1990.

The Free Fall Begins

In 1991, recyclers thought prices hit bottom, but it was just the tip of the iceberg compared to how far prices would plunge in 1993. Gluts of green glass, office wastepaper, and prime aluminum ingot, combined with a recession, export container problems, and a lack of demand for recycled materials, caused markets to fall in 1991.

Several large glass plants, including Anchor Glass Container Corp. (Tampa, FL), Miller Brewing Co. (Milwaukee, WI), and Owens-Brockway (Streator, IL), either stopped using or limited their use of green glass. Green glass processor prices ranged from zero to $7 per ton in 1991 compared with $8 to $23 in 1990.

Meanwhile, gluts of sorted white ledger, colored ledger, mixed office paper, and especially ONP, left many recyclers seeing red. In 1991, some processors had to paid as much as $50 per ton to get rid of their ONP.

The recession and the Persian Gulf War further exacerbated weak wastepaper markets. At the beginning of the year, shipping containers were scarce because of the Persian Gulf War. Often, material sat at docks rather than going overseas, cutting off one of the U.S.'s largest outlets for wastepaper. The recession also cut into wastepaper use; during the holiday season box orders were down because shoppers were spending less.

Prices of UBCs and high-density polyethylene fluctuated but ultimately remained low. The UBC toll price ranged from 32 cents to 50 cents, a 4 cent to 6 cent price drop from 1989.

Drop Continues

Markets remained weak in 1992, causing several plastics processors to go out of business. The advent of third-party processors for glass further pushed down glass prices, and aluminum remained weak as the cash-hungry Soviets flooded the market with materials. The only occasional bright spot was paper, but even that did not rise by any appreciable rate.

Low virgin resin prices made it difficult for recyclers to market recycled resin. As a result, DuPont sold its share of Plastics Recycling Alliance (Chicago). American Reclaiming (Houston) and Northeast Plastics (Lisbon Falls, ME) closed their recycling operations. Processor prices for clear PET ranged from 0.8 cents to 2 cents per pound; end-user prices ranged from 6 cents to 7 cents. Processors' prices for natural HDPE ranged from 0.6 cents to 1.6 cents per pound; end-user prices ranged from 6.4 cent to 7.8 cents per pound.

More and more glass manufacturers got out of the recycling business and began relying on third-party processors to supply them with recycled cullet. As a result, prices for glass dropped significantly. Processor prices ranged from $5 to $8 per ton for clear glass, $2.7 to $6 per ton for brown glass, and –$1.9 to $4.5 per ton for green glass. In the Mid-Atlantic region, green glass began to pile up.

UBC prices fluctuated throughout the year as cash-poor Soviets flooded the world market with aluminum. However, prices for aluminum slowly began to recover in late December.

Wastepaper prices gained some strength as more mills installed deinking capacity. CPO, white ledger, and office paper prices saw modest increases.

Although old corrugated container (OCC) prices were expected to take off because of new mill capacity, they never did. A major setback occurred when Stone Container's (Chicago) Seminole Kraft mill in Jacksonville, FL, abruptly halted operations when its roof collapsed on September 24, 1992.

The Bottom Falls Out

The beginning of 1993 started strong as prices for most commodities began to rise. However, by spring, prices for all recyclables began to falter. By early fall, prices for many commodities, especially paper and aluminum, had fallen to near historic lows.

Overall in 1993, glass, plastics, and steel markets remained relatively stable, with some price fluctuations, while markets for paper and aluminum weakened.

A number of factors caused these low prices, including a weak world economy, coupled with an oversupply of material on the world market. Germany's aggressive recycling law, an influx of aluminum from the former Soviet Union, and plans for China and several other Asian countries to begin producing their own plastics wreaked havoc on U.S. markets for recyclables.

Yet, 1993 was the year that recycling received a boost from the federal government. In October 1993, President Clinton signed an executive order directing every agency of the federal government to purchase printing and writing paper containing 20% post-consumer material by the end of 1994, and 30% post-consumer material by the end of 1998.

ONP began 1993 as the commodity to watch. After months of having a negative value, ONP had become a hot commodity. By late April, some end-users were paying as much as $23 to $35 per ton for this commodity.

By October, however, recyclers and paper brokers said they were lucky if they could move ONP or most other paper grades. Export markets dried up as the world economy remained weak and

Germany's aggressive recycling law continued to flood European and Asian markets with wastepaper.

On the domestic side, many mills started using just-in-time inventories, which allow mills to buy only enough wastepaper to run their operations for a limited amount of time. As a result, mills bought less wastepaper in 1993.

Aluminum prices also fell to near historic lows as an influx of aluminum from the former Soviet Union and an overcapacity of primary aluminum production in the U.S. kept aluminum prices lower than usual in 1993.

Citing an oversupply of aluminum on the world market, Alcoa reduced a quarter of its U.S. primary aluminum capacity. Reynolds Metals Company (Richmond, VA) reduced its primary aluminum production capacity by 9%. Despite the decreases in capacity, prices for UBCs remained weak in 1993.

By mid-November, the spot price for aluminum on the London Metal Exchange (LME) had fallen to 47 cents. Previously in 1993, the LME spot price had not fallen below 50 cents. As a result, prices in the U.S. for UBCs plummeted, with most areas paying below 20 cents per pound for UBCs.

Meanwhile, prices for plastics temporarily increased, only to remain flat for most of the year. Prices briefly rose in April and at the end of the summer in the Northeast, Mid-Atlantic, and East Central regions. Ironically, the lack of available plastics may have been related to 1992's glut. As communities found it more and more difficult to find markets for plastics, many dropped the material from their curbside collection programs.

High-density polyethylene (HDPE) recycling did not fare as well in 1993. While Graham Recycling Company (York, PA) announced plans to expand, North American Plastics Recycling (Fort Edwards, NY) and United Resource Recovery (Kenton, OH) closed their doors, citing poor market prices for the material.

In late October, North American was bought by World Class Film (Yonkers, NY), a manufacturer of recycled-content trash bags and film packaging. However, the company ceased processing HDPE.

Prices Take Off in 1994

Scarce supplies, new technology, and burgeoning processing capacity marked the recycling industry in 1994, signaling that the industry was moving from infancy to maturity. That move would have been huge news in any other year, but in 1994, the biggest recycling story was the unprecedented series of booming prices. How booming? End-user prices for aluminum used beverage cans (UBCs) climbed 120% between January and December, while end-user prices for natural high density polyethylene (HDPE) shot up 152% during that time and processor prices for old newspaper (ONP) skyrocketed 1,166%.

After two years of prices for old corrugated containers (OCC) rarely rising above $30 per ton, paper brokers and mills watched as the price shot up to more than $200 per ton during a three month period. Since 1992, industry experts had predicted OCC prices would climb as more capacity came on-line. Two years later, the price surged beyond their expectations.

OCC's price increase started quietly in the dead of winter; in January 1994, OCC showed some strength when unseasonably cold weather prevented trucks from making mill deliveries. Most industry experts said the price would fade once the snow and ice melted. However, the price remained strong through April, and gained increasing momentum in May.

By early July—a time when OCC prices are traditionally flat—the price soared to more than $200 per ton on the export market and more than $100 per ton on the domestic market.

When OCC prices gained momentum, the commodity did not rise alone. The effect of OCC's prices spilled over to other paper grades, including lower grades, such as old newspapers (ONP) and mixed paper. Mills began using these cheaper grades as substitutes for expensive OCC.

Prices for office wastepaper (OWP) also gained strength. Industry experts speculate that OWP may mirror OCC's rapid rise to the top, as more capacity to deink OWP comes on-line.

Reasons for the increase include a better worldwide economy, creating domestic and foreign demand; more capacity on-line but relatively flat collection rates; a higher demand for the finished product; and a market demand for post-consumer content.

Eventually other commodities experienced a price surge as well. By the end of the year, glass and plastics saw healthy price increases and UBC prices were at a four-year high.

Suddenly, instead of a glut of material, there was a glut of demand. Mills and other end-users found themselves scrambling to find enough feedstock to keep their plants running at capacity. And processors, who had been struggling for four years to stay out of the red, were suddenly receiving top dollar for their materials.

As recycling matures, the industry faces new challenges. Recyclers must find ways to feed demand by collecting additional high-quality material while keeping prices from skyrocketing too high. Otherwise, recycled feedstock may become more expensive to use than virgin feedstock, causing a glut of materials, and lowering prices, essentially proving the harsh adage that markets are cyclical.

EDITOR'S NOTE: All prices mentioned in this article are based on *Recycling Times* data. Prices are national averages tabulated to show movement over time.

****

HOW MRFs AND THEIR CLIENTS SHARE RISKS OF FLUCTUATING MARKETS

by *Tom Polk* and *Michael Knoll*

When this was written, Polk was founder and principal of Environmental Economics in Washington, D.C. As this book goes to press, he is project manager for recycling industries development for the Maryland Department of Economic and Employment Development. Knoll is a project coordinator with the New York City Department of Sanitation.

Picture this: your company has just been awarded a large, multi-year recycling services contract, and is looking forward to predictable cash flow, a good rate of return on capital investment and the development of a solid business relationship. And, as seen from the public recycling manager's point of view, looking forward to a stable outlet for materials, a budget for services to plan ahead with, and no need to rebid, negotiate, or change vendors for some time.

Sounds like the perfect business match—and it was for Seattle, Washington, until the bottom dropped out of the market for old newspaper (ONP) in 1989. At that point the city's Solid Waste Utility was into the second year of a five year deal with two private recyclers. Unfortunately, the drop in ONP prices meant that projected income fell dramatically. How do you stanch the losses?

Not Unique But Still Painful

The question illustrated by Seattle's experience is not unique—many material recovery facilities (MRFs) depend on revenue from the sale of recyclables to cover some element of their costs. But the situation is unlike those faced by other solid waste functions, and deserves some innovative answers.

In addition to the usual business risks of any venture, the processing and marketing of recyclables carries with it the added risk of price fluctuations in the secondary materials markets. This "market risk" must be added to those usually considered when the contractual business arrangements between the sponsoring agency and the private MRF operator are designed, so that the subsequent recycling operation will meet the needs of those concerned, even when market prices go crazy.

In some cases both the private MRF operator and public recycling agency might benefit by sharing the market risk through either sharing in the revenue or by using a price index of some type. This can do two things. First, it can help guarantee stable program operation (since drops in market prices won't destroy MRF financial health), and second, risk-sharing can allow the public sector to share in any price improvements. This arrangement won't eliminate risks associated with market prices but it does reduce, to a more acceptable level, the impact of market risk to a MRF operator.

Three commonly used contractual risk-sharing mechanisms are merchant-plant arrangement, revenue sharing, and indexing (Table 4.1).

The majority of MRFs in the U.S. are owned and operated by private firms. To these companies, market risk is a key part of their business. Known as merchant plants, they are developed independently of a government client. The merchant plant seeks feedstock from a number of sources and depends on the tipping fee and product revenue to be financially successful. A key feature of this arrangement is that the private operator markets the materials produced and keeps all revenue.

Clearly MRF operators have a tremendous degree of expertise, skill, and sophistication at knowing materials markets and predicting prices. For many years scrap metal and paper dealers have faced market risk every day. Despite their marketing acumen, however, many find prices exasperatingly hard to predict or control. Who could have anticipated the 1991 Gulf War, the effects of the latest recession, or the German Green Dot program?

Table 4.1 Highlights of Selected MRF Contract Arrangements

Merchant Plant	Index	Revenue Sharing
Risk wholly with operator	Agency assumes most or all (75–100%) of market risk	Agency assumes some (50–80%) market risk
Simple billing system	Index system must be created	Revenue share must be determined
Auditing simple	Auditing simple	Auditing complicated
Proprietary info not disclosed	Proprietary info not identified	Markets must be identified
Incentive to aggressively market	Incentive to aggressively market	Incentive to conservatively market
All revenue to operator	Revenue to operator but tipping fee may rise or fall	Revenue split between operator and agency

Some merchant MRFs can compensate for market swings by raising their tipping fee to make up for slumping revenues, or by limiting the types of loads they accept. Historically, private operators have done just this in the metal and paper segments of the industry. But nowadays many MRFs cannot fall back on that tradition, because they have long-term contracts that lock them into a tipping fee and prohibit them from refusing materials.

Locked-in tipping fees are more common partly because they are a typical characteristic of the increasingly popular full-service MRF contract. In such an arrangement, a private company designs, builds, and operates a plant on behalf of a public agency. As well as set tipping fees, longer contract periods are popular with jurisdictions because they make budgets more predictable, lend stability to recycling programs, and reduce the frequency of cumbersome bidding procedures.

An operator depending on product revenue can be squeezed between contractually fixed tipping fees and declining revenue. Take, for example, the case of Waste Management of Seattle, mentioned above.

When Waste Management signed its five year contract with Seattle's Solid Waste Utility, it agreed to collect, process, and market the city's recyclables in exchange for a fixed tipping fee and all income from the sale of recovered commodities. Within a year or two after the program started, market prices plunged and hammered the economics of the company's Seattle operation. "It was enough of a difference to make the operation unprofitable," says Waste Management's Don Kneass. Such painful experiences have convinced industry experts like Kneass that market prices are harder to predict than the Pacific storms that give Seattle its rainy reputation.

Market risk is a concern not just for MRF operators. It can also be a problem for their clients—the towns, cities, counties, and solid waste authorities that sponsor curbside recycling programs. For instance, a MRF unable to stay in the black might be tempted to discontinue service, restrict the materials it accepts, or close its doors completely. The threat or reality of a shutdown means that the public agency's basic mission of providing continuous waste management is likewise threatened.

Though less likely, prices might soar instead of fall, presenting another problem for the agency. Higher prices may be good for the operator, but public officials may fear being criticized for allowing private companies a revenue windfall.

Even with smaller drops in prices, an operator burned by falling product revenues in a previous contract is likely to raise the tipping fee when a renewal of the contract is negotiated. Sooner or later, MRF operators bidding on municipal contracts are likely to include in their bids a risk premium or insurance-like premium to cover the operator against hard-to-predict falls in market prices.

An Insurance Program?

How much of a proposed tipping fee is the risk premium? Usually the agency issuing the bid won't know, but in Seattle, it became clear. When Waste Management's contract came up for renewal, the company proposed two alternative tipping fees. Under one, the market risk would fall to the company, while under the other the risk would fall to the city. The difference: $6/ton.

Where public agencies and private operators collaborate under multiyear MRF contracts, volatile prices for recovered materials, reliance on revenues, and tipping fees which remain fixed can add up to a problem—a problem for the operator that can sooner or later become a problem for the recycling program that depends on the MRF. In such a setting both sides need a contractual relationship which clearly apportions the costs of market risk to either.

Contract Implications

The Merchant Plant Approach

In this type of arrangement, the public entity pays the MRF an operating fee, with the operator retaining all revenue from materials marketed. Financially, this means that the operating fee plus product revenues equals total revenue, with the incentive for the MRF operator to keep operating costs low and to effectively market the recyclables which are processed. Under a merchant plant agreement, the MRF has all market risk—the success of the venture depends on the accuracy of the operator in predicting market price changes.

A drawback of this arrangement is that if falling market prices make the operation unprofitable, the operator may lose the incentive to provide good service to the agency. If prices rise, the resulting windfall to the operator may draw criticism to the agency.

On other issues, the merchant plant approach fares better. Administratively, it is by far the simplest to understand and requires the least paperwork, factors which may keep costs down for both parties. Furthermore, the operation of the plant is the most flexible and allows the operator the greatest autonomy in making marketing decisions. Since the revenues go to the MRF, the changes in prices and specifications can be countered with changes in the level of processing and the resulting quality of the product, to name one example.

Revenue Sharing

In a contract with revenue sharing, the major difference from a merchant plant contract is that revenues are divided between the MRF and the public agency, reducing the market risk to the operator and increasing it to the municipality. In a typical contract with revenue-sharing provisions, the agency pays the MRF operator a fee to run both the plant and to market the recyclables recovered, but instead of keeping the product income, the operator turns 50% to 80% of it over to the agency.

An example of this type of contract is the construction and service agreement for the Staten Island Intermediate Processing Center in New York City, a 600 tpd MRF now on the drawing boards. The deal is a full-service contract—a single firm will handle design, construction, operation, and recyclables marketing. The City of New York will pay the operator a per-ton tipping fee to run the plant, with a minimum guaranteed weekly fee. The facility operator, a joint venture of Briarwood Construction Group and Resource Recovery Systems, Inc. (RRS), will manage the plant and market recovered products. New York City will receive 80% of the revenue from sales of recyclables. Briarwood/RRS will receive the other 20%, which can be regarded as a commission for marketing the recovered materials.

Revenue sharing is reportedly used at other MRFs; for example, in Monroe County, NY, where the agency share is 90%; Westchester County, NY (80%); Montgomery County, MD (75%); and Anaheim, CA (50%).

If commodity prices decline, a revenue-sharing arrangement shouldn't affect the fundamental economics of the MRF—unless the operator's bid is too dependent on product income—and operations can continue smoothly. In a similar way, both sides will see the agreement as reasonable, since revenue increases and decreases will be shared.

Finally, revenue-sharing may reduce the incentive to market the materials most efficiently, particularly if the operator's percentage of revenues is small. If operating costs, and some profit, are covered in the tip fee, the gain to be made from a 10% or 20% share of revenues may be less of a priority than with a merchant plant setup.

Administratively, a revenue-sharing contract presents significant issues. For one, the documentation required is substantial. The agency's auditors are apt to demand detailed records of individual shipments and sales of recovered products to be assured that the operator is reporting revenue fully. This will require the operator to give up some amount of proprietary information. If the facility's market outlets are named, the municipality or private competitors may use such information to the MRF's detriment. This has the further consequence of reducing the incentive for the plant to seek new and emerging markets.

Another consideration relates to the actual share of revenue. If the MRF has a high percentage of revenues, it may still shy away from aggressive marketing since its reward—a new market and better revenues—may be short-lived, as competitors or the municipality use the results of the market research to access that market for themselves.

With a low share to the MRF, the pressure to keep operating costs down and seek out relatively low-paying, conventional, markets will be great. Whatever the revenue split, it doesn't matter if the proprietary information contained in contract documents does not become public knowledge. The very possibility that it might may be enough to tip the balance toward a passive marketing strategy.

A second issue is that of the complicated bookkeeping needed with revenue-sharing. The MRF must prove its revenue amounts from all materials handled on a regular basis, and its detailed records must be reviewed by the municipal contract managers and auditors. Not only is there more paperwork for all concerned, but the information reported will probably require interpretation on a product-by-product basis. This has the potential to create misunderstandings, mistrust, and disputes.

The impact of the preceding issues on a contract's day-to-day workings at the plant level should also be considered. Additional time and energy spent on bookkeeping and clarifying individual product revenues will divert time away from actually managing the MRF more efficiently. The operator may not be motivated to give improved service if these dealings add too much to the contract's 'bare bones' service provisions. The plant's daily operation may also be affected with a small share of revenue to the plant.

Indexed Tip Fee

In this scenario, the tip fee fluctuates or is adjusted—according to an agreed index or indices representing the materials handled—at some regular interval, with the fee varying in the opposite direction of the price index. When prices are high, the tip fee is deflated; when revenues decline, it goes up. In this way, operating costs are covered in good times and bad, while increases in revenue are in a sense "shared," since the operator retains all income but must forego some amount of the tip fee when the index goes up.

The recently renewed contract between Seattle and Waste Management is a good example of an indexing scheme. The contract requires the city to pay the company an initial base tipping fee of $78 per ton to collect and process household recyclables. For each material on a list in the contract, the city makes an adjustment to its payment each month for market price changes.

The contract contains a fixed base price for each listed material. On the first business day of the month, the most recent "market price indicator"—a price report from a published source or telephone quote from a local market buyer of the material—is obtained. The city and the company agreed to use specific market indicators.

For each recyclable item, the difference between the market survey price and the contract's base price is multiplied by the number of tons of that material collected during the prior month. This gives the price adjustment for that material. If the material's price went down, the city adds the monthly adjustment to the payment it owes the company. If the price went up, the city deducts that amount.

Contract management, though not as simple as in the merchant plant model, is relatively straightforward. The only additional complexity is in the research and development of how the indexing system will work and justifying the choice of a specific index. Once in place, the monthly, quarterly, or annual adjustment is only a matter of a mathematical calculation; no confidential market information need be exchanged, and the danger of abuse is limited to the ability of the market indicators to be manipulated outside of normal business transactions.

In plant operations, an indexing mechanism provides an incentive to adapt processing activities to gain higher product prices. For example, if the MRF operator can obtain prices higher than the price adjustor—which in many cases is an average of buyers surveyed—the premium will be retained. As a further incentive to enhanced marketing endeavors, the indexing mechanism allows the identity of new market outlets to be kept confidential. The MRF reaps the full benefit of its efforts to secure better paying markets.

Putting Risk-Sharing into Practice

How can risk-sharing arrangements be set up? The answer to this question depends a great deal on local conditions and the nature of existing business relationships that MRFs and municipal authorities have. There are several key factors which should be considered in light of individual circumstances.

First and foremost, determine the amount of risk you want. For the MRF, this means examining your current market outlets and potential to obtain new ones. The MRF operator should also factor in the cost of running and maintaining the plant.

The municipality should decide how much of the market risk it wishes to assume. If it is considering placing market risk on the operator, it should ask how much risk premium prospective contractors are likely to charge and decide how much of a risk premium is affordable. The more risk the operator takes, the higher the net fee is likely to be. The public agency must weigh whether it is willing to purchase financial stability for the operator by assuming more market risk. With a risk-sharing arrangement the risk premium the agency pays will be lower, but the budget impact is uncertain. If, for example, an index is used, the tipping fee may increase and additional funds to continue the contract would have to be found—a prospect which public officials must recognize.

If the agency wants to assume some of the market risk, the next question is whether an index-based contract or a revenue sharing arrangement is the best for a given situation. Neither system may be appropriate for all localities. The pros and cons already discussed may shift the balance in favor of revenue-sharing or indexing due to any number of program- or site-specific conditions. Once this decision has been made, the factors to consider become more specific and vary with each approach.

Regardless of the business arrangement, it is advisable to communicate with, and solicit information from potential partners. By initiating such a discussion, many potential misunderstandings and disagreements may be forestalled.

Index Contracts

For index contracts, consider the following when looking at what market indicators to use:

- Use of price tables, as published by most trade journals, where "nominal" or "transaction" prices are quoted, versus a true index, in which prices are converted into decimal numbers and fluctuate based on a percentage change in actual prices (the U.S. Bureau of Labor Statistics—BLS—publishes the Consumer Price Index and the Producer Price Index in this manner);
- For price tables, sometimes a price range is given instead of a single number—using the midpoint is one way to keep this from introducing ambiguity into the indexing formula;
- What is the frequency of the indicator? The shorter the reporting period, the smaller the market surveyed;
- What is the lag time between obtaining prices and publication?;
- What mechanism can be used if the price indicator is not available (both for a once-only occurrence and when a particular index changes drastically or becomes extinct)?;

- How large a market region will be covered?—local, regional or national prices.

Care should be taken to make sure that the index is reflective of actual prices. As any individual operator's market prices and specifications will not be identical to the price adjustor used in the contract, no market indicator is going to be an exact fit. In many cases the index is a proxy for the movements of a number of product specifications. It doesn't matter as long as the index varies consistently. To help determine what price indicators may be suitable, the agency preparing the bid should find out what MRF operators think of existing indices. The local MRF industry may have experience with a price adjustor that could be valuable to the municipality preparing the contract.

Another potential pitfall is that of the index lagging too far behind prices in the market. A price adjustor that doesn't adjust is not much good. Daily, weekly, or monthly surges or retreats may be lost in an average for a week, month, or quarter, with cash-flow problems on the one hand if operators' lower prices are not reflected in a timely way. To deal with this problem, an index that is available over a shorter time period may be used, or price adjustments can be made retroactively to the previous time period.

If a number of recyclable materials are involved, some means of combining a number of indices should be considered. This could take the form of assigning weights to each recyclable, based on the proportion of each going through the MRF, and arriving at a "market basket" adjustor.

Revenue-Sharing Contracts

Developing revenue-sharing agreements requires fewer initial considerations than an index arrangement. These include determining the percentage share for operator and public agency. The impact of allocating risk with higher or lower percentages for each party has to be acknowledged. A second item which may be part of the basic contract is whether there should be floor prices below which one party would assume all losses.

Several more detailed issues arise when contemplating the mechanics of administering the revenue-sharing provisions in a contract. For example, revenues and operating costs may be intermeshed. The freight cost may be covered on some items, or a central marketing department may allow some plants to make some sales decisions, resulting in different prices paid; a broker's fee may be part of one commodity price and not others, etc.

To prevent the difficulties which could flow from trying to determine what's revenue and what's an operating cost, the bid should identify what counts and what doesn't, beforehand. The balance to be struck should be that of allowing the maximum flexibility in the method of sales while ensuring that the benefit of this tool for increased efficiency is part of the revenue split.

Another concern to be addressed when developing the contract is that of the nature and extent of recordkeeping and market information that the agency will require to verify the amount of revenue to be shared. Requiring minimal amounts of documentation to begin with is helpful. Then the agency should make provisions for penalizing noncompliance with these requirements. In addition, the agency should closely examine how it can best protect any proprietary information that may be submitted as part of the revenue-sharing paperwork and how it could reduce the exposure of this kind of information.

Among the thousands of curbside recycling programs, only a few use an index. Those that do, however, are some of the most successful and largest programs in the country, including Philadelphia and New York City, as well an interesting hybrid system in Harford County, MD.

Philadelphia

The City of Philadelphia has used some form of an indexed tip fee for over five years, and city officials have revised and refined bids at every opportunity. In current contracts, six separate indices are used for commingled containers, and weighted as follows: clear glass, 36%; brown glass, 18%; green glass, 18%; tin and bi-metal cans, 26%; and aluminum, 2%. A single index is used for ONP, which is calculated and accounted for separately. For bottles, cans, and ONP, prices are adjusted on a quarterly basis. Philadelphia's contracts share only part of the market risk: the contracts limit the city's risk to 75% of the market risk. The indexing formula allows the operator to pass along to the city only this portion of changes in prices.

New York City

In New York City, indexing came into effect in 1990, but for paper (ONP, corrugated cardboard, and magazines) only—other contracts were either on a merchant plant or revenue-sharing basis. New York City uses a decimal-based index rather than industry trade journal prices. The city uses one of the commodities tracked by the Producer Price Index (PPI) published by the BLS in the Department of Labor. The PPI measures thousands of separate commodities and the BLS uses a selected number to evaluate the future direction of the economy. For city contracts, only one of these indices is used. Termed "#1 News, Northeast Region" by the BLS, it includes several ONP and groundwood grades of wastepaper. As with Philadelphia, refinements have been made over the course of time. One change was to reduce the period for adjusting prices from two months to one; in this way price changes are counted every month, rather than averaging two months together and using the same number for both months. Unlike Philadelphia, NYC allows 100% of index changes to be included in adjustment calculations. Meanwhile, the city is continuing to use both merchant plant contracts and revenue-sharing agreements. Contracts for commingled recyclables (glass, tin, aluminum cans and foil, PET and HDPE) continue to use the merchant plant arrangement, while the new Staten Island facility will use revenue-sharing.

Harford County, Maryland

The most unique risk sharing arrangement we've seen is described by Bob Ernst, recycling coordinator for Harford County, near Baltimore. The county sends its recyclables to a Browning-Ferris Industries (BFI) plant in Elkridge, MD, and pays the company by the ton. BFI, in exchange, rebates 100% of the revenues from the materials processed.

Since BFI's MRF is a merchant plant serving several Maryland counties, the MRF mixes all its customers' material together during processing. Each agency's recyclables may have a different composition, too, so it could be difficult for the MRF operator to track or estimate Harford County's revenue separately from the revenue from the material of other customers.

So instead of basing its revenue rebate on actual shipping records and invoices for the sale of recovered products, BFI bases its rebate on an estimate. The estimate for a given material equals the number of tons of material delivered to the plant, multiplied by that material's proportion of the feedstock in the most recent semiannual sorting study, times the material's published market price.

Prices for materials recovered at the curb will probably continue to fluctuate. Establishing effective and efficient recycling requires adapting to this market risk.

5 RECYCLING COLLECTION AND VEHICLES

(There are a variety of methods of collecting recyclables—separate collection and separate vehicle, co-collection with the regular trash and use of specialized vehicle, or blue bags, with the recyclables placed in colored bags and collected with the rest of the trash. The articles in this chapter discuss these various methods.)

CO-COLLECTION TRUCKS

By *Michael G. Malloy*, contributing editor to *Waste Age,* Washington, D.C.

The co-collection vehicle, which accepts both recyclables and refuse into different compartments of the same truck, is a concept that is catching on in rural settings, medium-sized areas, and even in some large cities nationwide.

While generally seen as best suited to rural applications because of the distances involved, co-collection vehicles are touted by their backers as big potential money savers in equipment, labor, fuel, and insurance costs, as well as for their potential environmental benefits in eliminating the need for a second truck.

The use of co-collection trucks in rural areas can make sense because sending both a refuse and a separate recycling truck is not generally economical over a wide area. But while even their proponents say the vehicles are not likely to be well suited for use in populated urban zones, some smaller to mid-sized localities are proving co-collection's benefits.

A look at several locales using co-collectors—and at the companies that are making and/or that modify them—shows the wide variety of styles and features with which innovative haulers and municipalities are experimenting.

Most of the programs using co-collection trucks have custom-built the vehicle to get the various compartments to fill at the same time—either after careful planning as to when the separate collection areas of the body fill up, or after an initial trial-and-error period.

Rural Applications a Natural

Waste Stream Management (WSM, Potsdam, NY) hauls about half the trash in New York state's St. Lawrence County in custom-designed co-collecting vehicles. At 2,842 square miles, the county is the nation's largest in area east of the Mississippi River.

WSM began co-collecting last year, and uses one-man, stand-up, right-hand drive trucks to serve its rural customers. Its seven co-collectors have three recycling bins on the side, which are designed with 5.6 yards for paper, 2.7 yards for commingled bottles, cans, and plastics, and another 2.7 yards "for whichever bin fills up first," says Skip Bisnett, WSM's vice president.

The low-profile, modified Internationals have 8-yard, side-loaded Burro refuse packers, and include lifters for 90-gallon Zarn refuse carts. WSM had been pulling separated recycling trailers behind its trucks, but concerns over insurance and commercial driver's license certifications prompted the company to think of an alternative, Bisnett says.

"We can put 300 to 400 stops on a truck per day," he says, adding that some of his drivers log as many as 200 miles per day, some traveling 5 or 6 miles between stops. WSM has also sold a modified co-collector to another New York hauler.

Ontario Town Tries Wet/Dry

Co-collection programs are not limited to the U.S. A city not far from Toronto has been running a co-collection pilot project on about 900 homes for the past three years, using two different pickup methods.

Guelph, Ontario, a city of 70,000, has been running a "wet/dry" in both two- and three-bin configurations, says Janet Laird, the city's waste management coordinator. To get better waste stream diversion,

Guelph has decided to go with a two-bag system, in which residents separate their trash into dry goods (clean recyclables and dry refuse such as old sneakers) and wet compostables.

The city is scheduled to open a huge, $36 million (Canadian) integrated waste management facility in late 1995 that would be "the first of its kind in North America," says Laird. It would encompass a materials recovery facility, organic waste processing and composting, household hazardous waste, a drop-off, and collection bunkers for such items as tires, scrap metals, and white goods.

The facility will open to residential use first, says Laird, likely in the fall of 1995, and then to commercial and industrial waste as early as 1996. With the facility's commissioning, no waste would go directly to the landfill, Laid says; landfills would only take the residues left over from the waste processing.

Guelph is still mulling over exactly which truck configuration it will go with to implement the plan, but wants to use an efficient, one-driver system.

The city is looking at horizontal and vertical lifts, as well as automated and semiautomated systems, Laird says, adding that so far, the city has received "about eight proposals" from various body manufacturers.

Colorado Cities Co-Collect

Several factors have converged to make co-collection systems attractive to some of Colorado's cities. With recycling in the state generally not mandated, and markets for them often as far away as West Coast ports, recyclers in the state have their work cut out for them.

"I think we're one of the first companies doing co-collection in a suburban area," says Dick Ross, general manager of Best Trash, a relatively new hauler in the Denver area, which began using the system in May 1992. Although Best Trash had been in business for just a little over a year, "we were already convinced that a one-truck system made sense," Ross says. "It was just more efficient."

The cost comparisons between one- and two-truck systems are a bit difficult for his company, Ross says, because it did not retire a recycling truck when instituting co-collection. But, he adds, the co-collector his company has been using has "saved us over half in terms of recycling cost."

The company uses three co-collection vehicles. Each truck, a modified International 4900 Series with 20-yard EZ-Pack body, has side openings for metal and glass, and five yards of capacity for newspapers.

Trial and Error

Virtually all locations employing some kind of co-collection scheme have had to experiment in order to come up with the most efficient design of how much space to allot to refuse and various recyclables. The city of Loveland, Colorado was no exception.

Loveland, a city of 30,000, uses co-collectors in tandem with a volume-based rate collection system. It instituted its overall plan as a result of spiraling costs, including such high-ticket items as workers' compensation and insurance.

With the institution of volume-based rates, recycling rates boomed, says Mick Mercer, Loveland's manager of streets and solid waste. Initially the test truck's recycling compartments were too small, while the refuse section was too large. Both had to be altered, he says.

"It would have been a fluke if our test truck was spec'ed correctly," Mercer says.

Loveland residents pay a flat $3.40-per-month service fee, plus 75 cents for each 30-gallon trash bag they use. More than 90% of the residents participate in the curbside program, Mercer says, adding that recycling saves the average household about $1.50 per month. "We don't rely on people's environmental awareness . . . We go after their pocketbooks," he says.

The city has had its full force since January 1993: five co-collection vehicles, along with two older trucks that it uses for yard-waste pickup. This compares to what it would have otherwise needed, Mercer said: five one-man, side-loading refuse trucks, three recycling trucks, and two yard-waste collectors. (Yard-waste collection in 95-gallon Schaefer carts is optional to residents at $2 per month.)

The labor savings are basically a wash, Mercer says, but by using co-collection vehicles Loveland saves the capital and maintenance costs of three trucks, which run about $120,000 each. The city saves about $100,000 a year, he adds.

Environmental Benefits Noted

Loveland's trucks have Crane Carrier chassis, while the EZ-Pack bodies were modified and the co-collecting recycling bins added by May Manufacturing (Arvada, CO).

May, which gave Loveland its first modified unit in order to let the city experiment with the system, is one of several companies that design and modify standard waste vehicles to turn them into co-collectors. May's systems do not compact the side-loaded recyclables but do have lift arms on the sides to pick up carts.

Jim McMahon, May's marketing director, sees a good future for co-collection systems: "Co-collection is where recycling has to go," he says. "The cost of separate collection will drive it that way."

Using one vehicle instead of two can have positive environmental benefits, McMahon says, especially in a state like Colorado, where Denver's air quality is routinely on national lists of the most polluted urban air basins.

In a competitive suburban hauling scenario, there could be five haulers serving a single street. If three of these haulers also have recycling trucks, that makes a total of eight trucks on one street, McMahon notes.

"In Colorado, there is no environmental benefit under that scenario," McMahon says. "And the cost of fuel outweighs any benefits that would have been gained by recycling."

McMahon cites the city of Seattle, Washington—whose recycling program he helped put together in the mid-'70s—as one example of a city where co-collection would not be effective. Residents put out so many recyclables in such urban areas that "separate collection [there] makes sense."

But, he adds, "co-collection is coming on slow," and especially in more rural areas, "there's no question that co-collection is the way to go."

May has retrofitted the co-collection bodies of trucks in seven states, including an 18-yard Dempster in Port Jervis, NY, a 20-yard Loadmaster in Telluride, CO; a 20-yard Leach in Hampstead, MD, and an 18-yard EZ-Pack Apollo in Smithfield, NC.

Recyclable Compaction Offered

Another company that makes co-collection bodies, G&H Manufacturing (Arlington, TX), uses compactors in its recycling compartments.

G&H makes two- and three-compartment, rear-loaded bodies, custom designing them to buyer specifications.

A typical setup for a three-section truck has newspapers and commingled goods flanking either side of the refuse section. Each section has its own compactor and tailgate, and cannot be contaminated by the other streams, says Greg Hunt, G&H's general manager.

The two-bin configuration can be used in a variety of pickup plans; New York City ordered 30 trucks with plans to experiment with various streams.

Hunt says that compacting recyclables does not cause much more glass breakage than does dumping noncompacted loads, and adds that newspapers could be compacted in at about 800–850 lb/cu.yd., much better than the 300 lb/cu.yd. that G&H had anticipated.

G&H, which has also received co-collection orders from trash giants Waste Management, Inc., and Browning-Ferris Industries, makes co-collection bodies ranging from 20- to 35-yard capacities. They can be put on virtually any chassis a customer wants, Hunt says.

Co-Collect in My Back Yard

Shaker Heights, Ohio, a Cleveland suburb of about 35,000, drew up its co-collection plan based on the backyard trash pickup it provides to its residents.

The city received $125,000 from the Ohio Department of Natural Resources—and spent about $300,000—retrofitting its fleet of 20-yd. Heil-body packers, which include five bins on the side for separated recyclables. The system has worked well, says Randy DeVaul, director of public services, who credits his staff for coming up with the co-collection plan. Retrofits to the city's trucks were done by both May Manufacturing and Power Brake and Equipment Company (Cleveland).

The main reason the city went to co-collection was "to put the money into equipment rather than [into] labor costs," DeVaul says, though he adds, "I'll have to admit, if we had curbside service we might have gone a different route."

So, with the wide variety of co-collection trucks and pickup schemes emerging around the country (and beyond), there may not be *the* truck of the future, but chances are that in many locales and applications, there will be a truck of the future.

****

CO-COLLECTION: IS IT FOR YOU?

By *Jennifer A. Goff*, managing editor of *Recycling Times*, Washington, D.C.

"It would not be a stretch to see 25% of curbside collection . . . being [served] using co-collection [methods] within three to four years," predicted Ron Perkins, director of recycling operations with the American Plastics Council (Washington, D.C.) [in 1992]. But with high-tech co-collection vehicles, blue bags, and even modified units, the debate continues as to which co-collection system answers three important questions: Which method is the most expedient? Which technique produces quality, contaminant-free recyclables? And, bottom-line, which system is the most cost-effective?

Realizing the Problems

Today, when a hauler is weighing the advantages and disadvantages of "traditional" versus co-collection service, the first items that need to be considered are the obstacles that the route presents—or may present—to a particular method of collection.

Questions to ask:

- Location—Is it a rural, suburban, or urban route?
- Materials recovery facility (MRF) proximity—Is the MRF close to the landfill or transfer station?
- Wages—Are wage rates high?

Once you have the answers to these questions, it will be easier to customize your collection service to suit your route.

"There's no question that if you're on a rural route, you should definitely be co-collecting," asserts Jim McMahon, [then-]marketing director for May Manufacturing (Arvada, CO). "In a rural area, it makes more sense to have just one truck out there," concurs Jonathan Burgiel, director of materials recovery for R.W. Beck and Associates (Orlando).

The advantages of co-collection in a rural area are numerous, according to McMahon, Burgiel, and Perkins. For one thing, "you're not sending two trucks down the street tearing up the roadway," Perkins says. You're also "reducing the amount of fuel usage, and when you really look at it, it's saving all the driving time of the two trucks."

As for urban settings, the advantages of co-collection require more careful analysis. Factors such as wage rates and tight streets may affect not only the cost-effectiveness of the system, but the overall service as well. "I wouldn't recommend [co-collection] for a major municipality [that] can send out a separate truck [to accommodate a high volume of recyclables]," McMahon admits.

On the other hand, most co-collection systems require only one or two employees to both operate the truck *and* collect the refuse and recyclables. Separate collection requires not only two trucks, but generally more personnel. Consequently, "in urban areas, you may want to put more money in the [co-collection] equipment if you have higher wage rates," Burgiel says.

Another logistical consideration for urban areas is the longer length of most co-collection vehicles. Most cities have narrow streets and tight corners that may be hard to manipulate. Still, with a little planning, this problem can be overcome as well. "Take a look at the chassis and attempt to compensate for a longer vehicle," McMahon says. "You need to make sure you have the same turning radius."

We're in the Money

In some ways, it is easier to determine the cost-effectiveness of co-collection for rural and urban routes, simply because factors such as driving time and wage rates are relatively easy to identify and measure.

In terms of the cost of collection, the suburban route presents a more complex set of issues and requires a more detailed investigation into the potential benefits of one system over another.

Granted, co-collection offers the aforementioned advantages such as the potential for reduced wages, reduced fuel costs, etc. At the same time, co-collection also means more time out on the route because of the time it takes to collect both the recyclables and refuse, as well as investing more money in equipment.

As McMahon points out, haulers really have to do some comparison shopping. "You have to look at the cost of our unit [the Western Curbside Collector]; a half-hour extra on the route each day; and the cost of cab conversion, and compare it to the cost of separate collection."

R.W. Beck conducted a comprehensive study that addressed collection costs of several co-collection pilot programs in South Florida.

Not surprisingly, the specific results, in terms of the cost-effectiveness of the individual systems, varied. Overall, however, "The bottom line was that the co-collection systems were [generally] 13–15% more cost-effective," Burgiel says.

According to the study, "principal factors which affect the cost in the analyses when comparing total cost per household per month were found to be:

- Truck capacity by material;
- Number of employees used per truck and their salaries;
- Cycle time during collection;
- Household participation rate;
- Amount set out per household by material; and
- Off-route time."

Unproductive, off-route time is a critical issue when considering the economics of co-collection. "We try to oversize the recycling compartment in the truck . . . so that it's the trash, not the recyclables, that drives that truck off the route," McMahon explains.

Skeptics of co-collection are particularly concerned about plastics because of the amount of room that plastics tend to require in the truck. But in the Lake Worth pilot program, plastics were "collected in an Oshkosh collection vehicle equipped with a 17-cubic-yard side-loading refuse compaction unit," according to the R.W. Beck study. "As it turned out . . . the refuse body filled up just about the same time as the recyclables section filled up," Perkins says. "I'll say with a plastics compactor, you'll never have to go off route [specifically because of plastics] because it will hold 300–400 pounds of plastic."

Another related, off-route problem is the location of the MRF. "You lose the economies of one-stop dumping if the MRF is not close to the landfill," Burgiel says. However, "if you have a longer-term view, and you can locate the MRF next to the waste disposal facility, [co-collection] really makes sense."

So, What's the Problem?

Based on the studies that have been conducted so far, co-collection would seem to be the answer for haulers who are trying to cut costs, as well as the solution for those trying to make recycling work when markets are low. So why isn't it catching on?

"The obstacles are resistance to try something new," McMahon explains. "We talk to cities and private haulers all the time . . . Although they don't like the *cost* of separate collection . . . you have 'rules of thumb' in separate collection. In co-collection, the trick to building the truck is to size all of the compartments so that they fill simultaneously. That requires really thinking about the routes ahead of time. People just want to order a truck."

Another reason, according to Perkins, is that recycling really only started to boom in the late 1980s. "There's a lot of equipment out there that's still relatively new. When that equipment wears out . . . [co-collection] will definitely catch on."

****

AM I BLUE?: A REPORT ON BLUE BAG RECYCLING PROGRAMS

By *John T. Aquino* and *Kathleen M. White*
Aquino is Editor-in-Chief/Publishing Director and White is senior editor for *Waste Age* Publications, Washington, D.C.

When New York City announced it was going to convert its recycling program to a blue bag system, it also declared it was going to save approximately $11 million in doing so. The numerous articles citing the city's cost savings that followed the announcement prompted communities around the country to investigate, request, or even demand blue bag recycling for themselves.

In fact, the litany of the advantages of the blue bag system—a method of collecting usually commingled recyclables in blue-colored bags, often alongside municipal solid waste (MSW)—have been repeated from Delaware westward: blue box containers don't have to be purchased (the $11-million New York figure represents the cost of bins); in many cases the recyclables and nonrecyclables are collected by one truck, and as a result, the city does not have to run dual fleets; the use of bags saves on collection time since the step of bringing the empty bin back to the curb is eliminated; for the homeowners the bags are easier to store than bins; and, finally, the bags keep recyclables more contaminant-free.

Yet in February 1992, after a test run, the city of Houston announced its intention to concentrate on bins, citing the expense and low participation rate of the blue bag program as prime factors. In addition, the long-awaited Chicago program is stuck in its pilot phase. Even programs that are regarded as successful have reported problems—Omaha, Nebraska's, participation rate, for example, is only

50–60%. And environmentalists note that since not all bags in use are recycled, they are a one-use item that contributes to the waste stream.

To provide an overview of the status of blue bag recycling in this country, *Waste Age* contacted spokespeople for some of the major programs in the U.S.

Pittsburgh

Although one-bag recycling actually started in California in the 1970s, Pittsburgh pioneered the use of blue bags, says Anna Maria Lozer, Pittsburgh's assistant recycling coordinator. "The program serves 370,000 residents," Lozer says, "and we haven't had any problems. Things have been running smoothly."

According to Lozer, the advantages of the blue bag program are "the convenience to residents and the ease in collection. There's just one trip to the curb for both hauler and homeowner, and the homeowners can get their bags free. One of the major supermarket chains in Pittsburgh, Giant Eagle, switched to grocery bags that are blue and other supermarkets have followed suit."

Lozer cites a very high participation rate of 82% for the blue bag program. In April 1993, the program was generating 75 tons of recyclables daily. Plastic, metal, and glass are commingled in one blue bag, while old newspapers (ONP) are separated in another. The bags are collected once a week—at a different time than MSW collection. After stopping at a transfer station, all materials, with the exception of ONP, are taken to a materials recovery facility (MRF) run by Browning-Ferris Industries (BFI, Houston) in Carnegie, PA. The ONP, meanwhile, is shipped to a deinking mill in Ontario, Canada.

BFI began processing the recyclables collected in Pittsburgh's blue bag program in early 1993 at a cost of $31.60 per ton, according to Maribeth Rizzuto, recycling coordinator for the city [when this article was first written and, as this article is being revised, currently with the Steel Recycling Institute], who has been with the program since its inception in September 1990. At the time the program started, Chambers Development Company (Pittsburgh) had the exclusive processing contract with the city and was actually paying the municipality $2.18 for every ton of recyclables it received. In 1992, the company was charging $3.77 per ton to process the recyclables. Chambers held the contract for three years until BFI took over in January 1993, outbidding the hometown company on the cost of processing. This 800% increase in processing costs reflects costs that were only realized once the blue bag program was underway, says Jamie Hill, general manager of Chambers' American Recycling Division. In addition, plummeting markets and higher labor costs drove up processing prices, she says. "Chambers has provided the city of Pittsburgh with exceptionally low prices for recycling over the last few years," she said upon the termination of the company's contract. "With this experience, we're learning the full cost of recycling. Right now, the costs are outweighing the benefits," Hill said.

Facing higher processing costs, the city has been forced to pass some of these added expenses on to the public and, as a result, has come under fire from some residents to drop the blue bag program. According to Rizzuto, however, the problem is not in the bags but in the nature of recycling itself. "Everyone knows that the most expensive part of recycling is collection," she says. "Using blue bags keeps costs manageable because collection is easier. It's not the purest form of recycling, but you have to think in [cost-effective] terms."

Omaha, Nebraska

Omaha's blue bag program was initiated in 1991. A prime reason for opting for blue bags, says Bob Sink, manager for environmental quality, was cost savings. "The program cost $400,000 last year. We estimate that, for our 100,000 households, the expense for citywide curbside collection would be $3 million a year. That's based on the experience of a neighboring municipality that was paying $2-plus per household for a bin program."

Participation rates are "steadily improving," and, according to Sink, "as the program grows, so will the cost." He estimates the expense is now "somewhere between $400,000 and $800,000." A substantial part of the cost savings, Sink notes, is that the city co-collects the blue bags of recyclables along with MSW, and in doing so, has not had to pay for a second fleet of recycling collection vehicles or the workers required to operate them.

Contamination has not been a problem for Omaha's program, so far, and Sink claims that, to date, manufacturers have not rejected a single load of recyclables. In addition, the city's collected newspaper, he says, "bypass[es] the quality sort in Portland [OR]" on its way to a Weyerhauser mill in Longview, WA. "It's rated #8 Grade A, –0.52% rejects or out-throws," Sink says. He also cites the survival rate of blue bags in the collection trucks at 90%.

As for the future of blue bag recycling in Omaha, Sink is cautious, noting that "we are locked into a contract for the current system that runs through 1995. We are just beginning a study for a 20-year plan that will meet the Nebraska legislature's goals of 25% reduction by 1996, 40% by 1999, and 50% by 2002." In 1992, says Sink, Omaha had a 6% reduction in the waste stream as a result of the blue bag program and a 20% reduction in the waste stream overall.

In giving a general evaluation of the blue bag program in Omaha though, Sink is cautious. "It [the blue bag program] was off to a slow start, but it's getting better," he says. "It's a voluntary program and demands active participation." [In March 1995, after rebidding its current contract, the city of Omaha elected to change its curbside recycling program from a blue bag to a bin system effective January 1, 1996.]

New York City

"The main reason for going blue bag was to give people an option," says Susan Glickman, director of marketing for the New York City Department of Sanitation. "Now New York residents have a choice: they can use a blue bag, a blue recycling container of the kind we had distributed, or a container of their own that has a recycling decal on it—we supply the decal."

Other reasons for converting to blue bags, she said, were the high cost of blue containers and the fact that many were being stolen. "We have 2.9 million households in New York City," says Glickman. "We can't afford to give out blue buckets [bins] to all those residents." Now residents who choose to recycle in blue bags must buy the bags, which are sold in neighborhood grocery stores and supermarkets. Packs of ten 13-gallon bags range from $1.49 to $1.79.

New York City's goal is a recycling rate of 42% by the year 2000. According to Glickman, "It's too soon to tell how the blue bag program is working. There have been no total studies yet." Currently, the city estimates a 45% participation rate for their recycling program. However, as Glickman points out, not all boroughs in the city have been converted to the blue bag system, and the participation rate reflects this. Glickman expects the rate to rise once the entire city is converted.

Nancy Wolf, a spokeswoman for Environmental Action Coalition (New York City), is critical of the blue bag program. "We feel it creates waste. Even though the [blue bag manufacturing] industry pledged that the bags would be recycled, they're [the city] left with 5 tons of blue bags that can't be recycled. We feel they violated the first order of solid waste, which is the reduction of waste."

On the streets of New York, Wolf says, people often tear bags open looking for deposit containers they can cash in on. "Also," she says, "blue bags were invented to put the recyclables and municipal solid waste in the same truck, and that's not what's being done in New York. These things [bags] get into the waste stream and drift on forever." Asked if she thought the city would drop the blue bag program, Wolf says, "It would take a major bit of political pressure to get this turned around."

Partly in response to concerns about blue bags adding to the waste stream, and out of frustration over the failure of the manufacturer of its blue bags to recycle 5 tons of the used bags, New York City in April 1993 announced its intention to introduce legislation requiring plastic trash bags to contain 10% recycled content by the end of 1993 and 30% by 1995. Mobil Chemical (Fairfax, VA), manufacturer of Hefty bags, told the city, according to Marcia Bystryn, recycling coordinator, that the 10 bales of bags they tested in November were too soiled to recycle.

Houston

After a blue bag pilot project that generated a great deal of press, the city of Houston announced in February 1992 that they would start favoring bins. The primary reason given for the decision at the time was the low participation rate of the blue bag collection method, according to Edward Chen, deputy assistant director for recycling. The participation rate for bags, said Chen, was 30% compared to 90% for the bin system.

The test project was conducted by BFI. George Elrod, environmental compliance manager for the Southwest region, says the test was both a success and a failure. "People enjoyed the convenience of blue bags," Elrod says. "But there was a problem in sorting. Bags were picked up at the same time as the regular trash—which, admittedly, can be an advantage of the blue bag program. But the newspapers and cans were commingled in the bags, and there was a contamination problem because consumers weren't washing the cans out. Overall, the sorting required makes it an extremely labor intensive operation."

In April 1993, Chen provided *Waste Age* with an update on bin and bag recycling programs in the city: "In Houston, 14,000 households are on a blue bag system with the city; 32,000 are in a bin program, and we are planning to add an additional 60,000 to the bin program this summer." That being said, Chen added that it was not a simple matter of saying "yes" to bins and "no" to bags. The high participation rate for bins, Chen says, is mostly attributable to the fact that these communities "came to us to have curbside. Our approach is that in the future, citizens who are really eager to have curbside recycling will get bins. We want to use the city's money for areas where we can be assured of high participation. For areas that are not as eager, we will use blue bags as well as drop-off and buyback centers. Houston is a really big city. And our program is a comprehensive one."

Elrod says that BFI is talking with the city about being involved in its recycling program, "and blue bags are on the list. One thing we've learned is that homeowner education—washing out cans, for example—has to be a large part of the program. We're even considering using a separate truck and evaluating the cost."

Mobile, Alabama

After 20 months, Mobile dropped its blue bag program in July 1992. "[The program] was extremely economical at first," says Bob Haskins, special projects coordinator for Keep Mobile Beautiful (and "recycling coordinator, when we had a recycling program"). But the cost-efficiency was due primarily to the fact that processing was handled by Goodwill Industries, which saw it as a state program to get employment for handicapped workers, he says. "It was basically a pilot program. We had average participation rates of 40–60%, serving 14,000 households, and it was growing. Over the 20 months, we processed 2.74 million lb. of paper, aluminum, and plastic. But as the volume increased, it got to be more than Goodwill could handle. So, its price went up. As a result we switched to Waste Management Inc., (Oak Brook, IL) for a six-month contract, and then to BFI for another six. The city council balked at the higher prices, and so when we realized they weren't going to put money in, we dropped the program," Haskins says.

There were complaints about the blue bag program, Haskins continues, mainly from residents who argued that the city was forcing them to buy the bags. The city had developed some options in response to this criticism, Haskins says, but they were never executed since the program was dropped. Collection occurred with one truck per route but on separate days, he adds.

Currently, the city of Mobile itself is offering drop-off and buyback centers "as a convenience to residents," Haskins says. "There is activity in the private sector. For the future, we have a new solid waste authority and are reviewing options for privatizing landfills, which may or may not include a recycling option." Asked if there's been an adverse reaction to the termination of the blue bag program, Haskins emphatically replies that there has been. The city's various recycling information phone lines get 300 recycling-related calls a month, and while not all of them are complaints, Haskins says the questions predominantly include, "Where can I resell my plastic?; Why aren't you doing more recycling?; [and] Where has my drop-box gone?"

In terms of future plans for the city's recycling program, Haskins is direct. "What I'd like," he says, is to "negotiate with a company, take it to a private MRF, and give it to them. Government doesn't have to be in the marketing end of recyclables. It's a tough market out there when you're trying to sell just 40 bales."

Chicago

Upon announcement of plans for the city's blue bag recycling program, Mayor Richard Daley had predicted that the program would be up and running in the summer of 1991. When this article was first written in 1993, Terry Levine, spokesman for the city's Department of Streets and Sanitation, said the program was now targeted for start-up in the fall of 1994. As this article is being revised in January 1995, the program has been scheduled to start up on December 4, 1995. WMX Technologies, Inc. [formerly Waste Management, Inc.] was awarded the contract to construct and operate four processing centers for the city, and at this writing construction is underway.

In May 1993, Levine cited as reasons for the postponement an exhaustive bid process and the usual governmental delays. In addition, a source familiar with the program said that the city's decision to create a Department of the Environment as a separate entity from the Department of Streets and Sanitation "fragmented everyone's attention." Meanwhile, debate over Chicago's decision to go with blue bags continues. Jo Patton, former executive director of the Chicago Recycling Coalition, was quoted by *Crain's Chicago Business* as saying, "The blue-bag program is bad." Patton estimated that of the 1.1 million tons of residential waste, only 2% would be recycled from the programs because of low participation and bags bursting inside the collection vehicles. Levine disagrees, citing a blue bag demonstration recycling project that showed that at least 90% of the blue bags survived in the collection trucks.

Myk Snider, senior consultant for Slack Brown & Myers, Inc. (Chicago), attests to this study, which was executed in 1991 when he was director of recycling for the Department of Streets and Sanitation. "We went to communities and asked them to switch from bins to bags for a while," Snider says. "We used our newest fleet of trucks to make sure the compaction was as tight as possible. Then we basically counted how many bags went in and how many came out."

What the demonstration results showed, Levine continues, was that "the blue bag program will be far less expensive than a bin program, if only because we won't have the cost of a second fleet of vehicles." Environment Commissioner Henry Henderson was quoted by the *Chicago Sun-Times* as comparing costs of $38 to $50 per ton for blue bag recycling compared to $150 a ton for regular curbside bin collection. The *Chicago Sun-Times* also noted that the city is issuing $41 million in municipal bonds to build two centers—down from the original plan of four. The original cost estimate for the four facilities had been $10–$15 million; however, the $42 million figure includes design work.

"The city still has complete confidence in the blue bag system," Levine says. "The 1991 blue bag test demonstrated that a blue bag program would be cost-effective for the city of Chicago and recover significant amounts of marketable recyclables," adds Snider.

Other Areas, Other Experiences

There are, of course, scores of other blue bag stories. Harford County, MD, initiated a blue bag program in 1992 to help achieve the state's goal of a 20% reduction in the waste stream by 1994; customers had a hard time finding the bags, and there was a contamination problem with the material being lumped together. Rumpke Recycling (Cincinnati) tested blue bags in 1992 as part of the regular trash pickup and concluded that since the recyclables were often contaminated or destroyed once the blue bag was compacted with other waste, a second or separate pickup was necessary.

Contamination of recyclables is primarily what forced the city of Pullman, Washington, to reconsider its blue bag program. In January of 1991 the city began a blue-bag pilot project that was designed to collect recyclables several days a week from approximately 25,000 residents. Problems arose when organizers realized that the blue bags were shredding in the collection vehicles and the recyclables becoming mixed in with regular garbage. "When they first started they were filling the trucks half full, which seemed to help a little bit with the contamination problems, but they had to make several trips," says Tim Davis, landfill operations manager for the Whitman County Sanitary Landfill, where the blue bags of recyclables and regular MSW were delivered. "Then they started to fill the trucks as full as they could, which only compounded the problem," he adds.

Pullman Disposal Service, Inc., the contracted hauler for the program, did adjust the compacting pressure on their trucks to try to decrease the breakage of the blue bags; however, the problem still persisted. Despite a good public participation rate, only about 30% of the recyclables collected were actually being recycled, Davis says. "The blue-bag program hasn't officially ended, but we aren't promoting it anymore," says Jeannie VanHouten, recycling coordinator for Whitman County. she says. "Citizens can still use the blue bags," VanHouten adds, "However, if they do choose to put their recyclables in them, there's a good chance they won't be recycled."

Although Pullman found the blue bag system to be an ineffective method of collection in their community, "We've heard that other communities have had no problems with contamination and have found blue bags to work quite well for them," VanHouten says. As noted, a blue bag demonstration project conducted by the city of Chicago in 1991 showed that contamination was not a significant problem. A similar test in Denver, Colorado, however, produced a different result. According to Danamarie Schmitt, recycling coordinator for Denver, results of a short-term pilot project testing the blue bag collection method in Denver last year showed the commingled approach produced too much contamination. Recycling officials eventually concluded that bin sorting collected the most marketable materials and was the most efficient operation for everyone.

Indeed, the conflicting outcomes and the variety of experiences reported by municipalities confirms that blue bag programs need to be evaluated on a case by case basis. Since the widespread implementation of this collection method is a fairly new phenomenon and the collection method itself varies—some programs co-collect blue bags with MSW and some do not, some programs separate materials by bag, for example—the results, for the most part, are still being measured. Whether the pros will outweigh the cons, or vice versa, still remains to be seen.

****

HDPE AND CURBSIDE RECYCLING

By *Jennifer A. Goff*, managing editor of *Recycling Times*, Washington, D.C.

Just the mention of "high-density polyethylene" (HDPE) and "processing" in the same sentence will make many materials recovery facility (MRF) operators shudder. Yet, increasing legislation and growing recovery rate mandates are turning some MRF operators' "nightmares" about HDPE collection into reality.

Still, adding "difficult" materials to curbside programs does not always have to be an ordeal. If the changeover process is assessed properly, many headaches can be avoided, according to Susan Smith, president of C2S2 Group, Inc (Seattle), a company that specializes in designing curbside programs.

"First, you want experience. *And*, you want a MRF that is flexible because you may need to change the way you are currently running your line," Smith says. "Some MRF operators have a specific way of operating . . . and sometimes [adding plastics] will throw their whole line out of whack. So flexibility is really the key."

But long before a MRF operator considers how the HDPE will affect the line, Smith recommends that he/she conduct a step-by-step, detailed appraisal of how the HDPE will impact overall collection, processing, and revenue.

We Shall Overcome

Fibers International (Bellevue, WA) operates a mid-sized MRF in south Seattle that has successfully managed to add HDPE to its list of recyclables.

"We negotiated with the cities to cover the cost of handling the stuff," says Greg Matheson, vice president of Fibers. "That is really essential. You have to find out how quickly your trucks are going to fill up with [the plastics]. The stuff is so darn bulky."

Adding plastics "item-by-item and step-by-step" was another key factor in the success of Fibers' program, according to Matheson. "The public is sometimes so confused about recycling plastics, you might end up getting a little of everything if you add them all at once."

Because of this problem, Fibers puts "emphasis on collection at the curb to prevent contamination," says Susan Robinson, project manager for Fibers.

"The drivers were trained [about contamination] when we first started the [HDPE collection] program," Robinson says. To combat curbside impurities, Fibers' drivers cover their 60,000-household circuit armed with "correction notices."

The notices are printed on brightly colored paper and are left at the curb when drivers notice contaminants in the materials. "Drivers demonstrate how to flatten the plastics, for instance, then leave a notice on it for the resident," explains Matheson.

While this kind of scrutiny takes a little more time, "the drivers who were most religious about [checking the materials] brought in the best loads, which saved us in the long run," Robinson adds.

Though careful collection helps prevent contamination, Matheson emphasizes extra safeguards in the MRF itself. "We use air classification equipment that we designed ourselves," he says. "Plastics and aluminum will blow off, but the glass will stay on. Then we do a positive sort, actually removing the HDPE from the belt. That is the only way you're going to get good quality."

But Matheson also warns against losing valuable operating time by storing vast volumes of material for processing in a huge baler. "It is ridiculous to take [a large-capacity baler] for wastepaper off-line. You want to have a system where you can drop HDPE right from the line and put it into a small-scale baling system." Fibers uses its GPI 1060 baler, which can process 1,100 pounds.

What About Markets?

Once a MRF operator has collected and processed the HDPE, where does it go? Market prices for HDPE vary throughout the U.S., but [when this was written in early 1994] the going rate was not exactly soaring.

Still, some see glimmers of future hope existing in the possibility of taxes on virgin resins, increased federal procurement of recycled products, government subsidies, and packaging standards—all issues waiting in the wings for support among members of Congress and local governments.

The bottom line is that when it comes to markets, recyclers will have to wait and see. A frustrating reality that MRF operators and industry professionals know only too well is that, though much legislation has been proposed and enacted to increase recycling rates, not much has been done to create markets for the end product.

In the interval, however, the emphasis should be on reducing collection and processing costs so that when better markets do emerge, operators are ready.

Though every MRF is bound to have unique concerns, Smith suggests the following general considerations and questions as a point of departure for any MRF operator who is considering adding HDPE to his/her curbside program:

- First, consider which types of HDPE you think you could take. Will you accept only clear HDPE? Tubs? Screw tops? Colored HDPE? Then, before committing to a specific type (or types), examine your local markets for the commodity.
- Then assess how much material you think you will amass. A good source to seek out might be city or county studies on plastics generation. By examining these studies, you can know what to expect locally—not only in the way of general volume, but also in terms of the volume of specific types of HDPE. After you have obtained the preliminary figures, calculate your annual volume, then divide that number by your number of operating days to get a daily volume. When you know how many cubic yards of material you will have to handle per day, divide that number by the number of trucks in your fleet. Keeping the capacity of each truck in mind, you should be able to roughly determine whether the truck(s) will have to go out a second time, or whether you will have to purchase new trucks with greater capacity.
- Also, take a look at the collection container capacity at each house. Will the addition of the HDPE containers make them overflow? If so, how will you deal with the excess? Will you instruct the public to bag the surplus separately? Or will you have to increase your current frequency of pickups?
- Then, look at the MRF itself. Think through the volume of material you expect to receive and translate that number into the number of containers *you* will need.
- Most of the containers will have to be picked off by hand. Smith estimates that one person can pull off one container each second throughout an eight-hour shift. (This estimate averages-in downtime.) Will the increased volume of material mean that another person should be added to the line?
- Yet another related consideration is that because the volume of plastics is so great compared to the weight, containers will fill up faster. You may need another person to move the containers around the floor.

- Contamination is also something to be considered. The above guidelines apply only to the loads of material that are acceptable. Pulling contaminants off the line will require more time, and may necessitate additional personnel as well—particularly if you plan to accept colored HDPE.
- Finally, educating the public is of paramount importance, since contamination and the economic viability of any recycling program are inextricably linked. Educational brochures not only need to be specific, comprehensible, and informative, they also need to be eye-catching and well-designed so that people will read them in the first place. Smith suggests hiring a professional either to put the piece together or, at the very least, to give input on the design.

****

MIXED PAPER—NEW TRICKS FOR AN OLD DOG

By *Chaz Miller*, manager of recycling programs for the Environmental Industry Associations, Washington, D.C.

Ten years ago paper recycling seemed to be so easy. Ten years ago Boy Scout troops had newspaper drives, less than five hundred curbside recycling programs collected newspaper in the U.S., and grocery store warehouses would bale corrugated boxes. In some office buildings employees were asked to place white letterhead and copier paper in desktop recycling containers, and computer rooms usually had a big bin to collect computer printout paper.

Today, newspaper is collected from the curbside in more than 5,000 communities across the U.S. and office paper recycling programs are common; an increasing number collect more than white paper.

Five years from now, curbside programs will collect magazines, corrugated boxes, and mixed residential paper. Office recycling programs will routinely collect an "office pack," or virtually all paper that is not food-contaminated paper.

Mixed paper is the grade of the future. But what is "mixed" paper? Who wants it? How will it be collected and processed? These questions need to be answered before mixed paper recycling can become a reality.

What Is Paper?

The paper used for magazines is a bleached, coated sheet of groundwood paper and is often one-third clay coating and filler by weight. It is vastly different from uncoated, bleached computer printout paper. And that paper is different from unbleached paperboard used to produce corrugated boxes. And that paper is different from the grey-colored recycled paperboard used to produce cereal boxes. Simply put, paper products vary tremendously. They have different fiber lengths and strength properties. They may or may not be colored, may or may not be coated, may or may not have been bleached.

Paper is the major contributor to solid waste by either weight or volume. In 1990, paper constituted 32% by weight and 31% by volume of waste sent to disposal, according to the U.S. Environmental Protection Agency.

Paper is also the major contributor to recycling. Of the 29 million tons of municipal solid waste recycled in 1990, the EPA says that 21 million tons were paper products.

Indeed, the American Forest and Paper Association (AFPA, Washington, D.C.) recently announced that it has met its 40% recovery goal for 1995. It now has a 50% recovery goal for the year 2000. Expected demand for recovered paper by U.S. manufacturers will drive the goal, according to AFPA's president Red Cavaney [who, as this was being revised, was chief executive officer of the American Plastics Association]. By the end of the decade, U.S. paper companies plan to invest nearly $10 billion in new manufacturing capacity to use recycled paper as a raw material. Currently, 145 new projects expanding consumption at U.S. mills are either up-and-running, under construction, or publicly announced, Cavaney says.

Increased mixed paper recycling is the key to the new goal. Mixed paper will grow more than any other grade of recovered paper, increasing from 4 million tons in 1992 to 11 million tons in 2000, according to an AFPA study. This is a 180% increase.

State recycling officials are equally aware of the importance of mixed paper in achieving higher recycling rates. "We've pulled off the best materials from the top of the materials barrel," says Tim Nolan, market development coordinator for the Minnesota Office of Waste Management (St. Paul, MN), "now we need to dig deeper to meet recycling goals. The paper streams will be more mixed, more contaminated, and of lesser quality."

What Is Mixed Paper?

"Mixed paper" is defined as "a mixture of various qualities of paper not limited as to type of packing or fiber content," in the 1993 Paperstock Guidelines published by the Institute of Scrap Recycling Industries (Washington, DC). The Guidelines define another 45 regular grades of wastepaper and list 32 specialty grades.

These different grades of wastepaper are used to produce different end products. Both technical and price considerations are factored into the final equation. A grade of paper that may make a good raw material for one type of paper, may not work nearly as well for other types of paper. For instance, corrugated boxes and brown paper bags are made with unbleached paper fibers. These "browns" cannot be used by mills that do not use chlorine bleach to produce an end product.

Using a different fiber than originally intended to produce the end product is called "fiber substitution." Examples include using an office pack instead of high grade white ledgers to make printing and writing paper, and using mixed paper in place of newsprint to make boxboard (the grayish multi-ply boxes used for cereals, shoes, and many other products). As supplies of old newspaper and old corrugated boxes tighten, mills that have used them as a raw material will be forced to use other grades of wastepaper, including mixed paper.

Markets First

Mixed paper has a wide variety of markets including printing and writing paper, towel and tissue paper, paperboard packaging, and non-paper uses. Of the paper markets, printing and writing paper has the highest end value, and packaging the lowest.

In addition, deinked market pulp is a dynamic new market. In a deinked market pulp mill, paper is deinked, made into a pulp, dewatered, and shipped to an end user, usually a printing and writing paper mill. American deinked market pulp capacity will grow from 424,000 tpy (tons per year) in 1992 to 1.7 million tpy in 1996, according to the AFPA.

The "next big wave," says Bill Moore, partner, Thompson Avant International (Atlanta), "is integrated printing and writing mills getting into the picture." Union Camp's (Wayne, NJ) paper mill in Franklin, Virginia, will deink sorted mixed office paper on-site and then use it as part of the raw material in producing printing and writing paper. "The end product will be 25% or more post-consumer content and will have comparable performance characteristics to a virgin product," says Dick Venditti, Union Camp's director of recycled fibers.

Recycled paperboard is also entering a period of expansion after a decade of decline in the 1980s. Including all the different product types defined by AFPA as recycled paperboard, 2.5 million additional tons of capacity will be added between 1992 and 1996. Some of this increase is in boxboard. Most of it, however, is in two paperboard grades—linerboard and corrugating medium—that traditionally use old corrugated containers, not mixed paper, as a raw material. European mills now use mixed paper as a raw material for linerboard and corrugated medium; American mills are beginning to, also.

Other mixed paper markets include roofing felt mills, insulators, fuel pellets, and exports. Mixed paper supplied 931,753 tons, or 15% of total wastepaper exports during the 12-month period ending in October 1993, according to the U.S. Department of Commerce. Although wastepaper exports as a whole were down by 550,000 tons for the first 10 months of 1993 compared to same period in 1992, mixed paper exports are up by 55,000 tons. Canada is the largest export market for American mixed wastepaper.

Market Specifications

Mills using wastepaper have very specific requirements which dictate the "mix" of paper they will buy. Wastepaper dealers and brokers and mill wastepaper buyers stress that collectors and processors have the needs of a specific end-user in mind, and do not produce a "generic" bale of mixed paper in the hope that someone wants it.

"Sorted mixed office grade is the wave of the future," according to Union Camp's Venditti. Paper brokers and industry analysts tend to agree with Venditti, but they also will say that the definition of sorted mixed office paper is confusing. "I know it when I see it," says Bill Moore, who adds "it is a grade in transition whose definition is very mill specific."

Recycled-content printing and writing paper mills have traditionally used pre-consumer pulp substitutes and post-consumer white high-grade office paper. For this market, the paper will have to be well sorted to meet mill specifications. Mixed paper will have to be sorted to pull out the browns and groundwood (including newspaper). Sorting can be expensive. Deinked market pulp mills will have similar specifications, since they generally make pulp for printing and writing paper mills.

Towel and tissue paper mills use varying amounts of recycled paper. As a result, many of these mills have a great deal of experience with recycled content, especially in making institutional products. Specifications for tissue mills will not be as tight as those for the printing and writing paper mills. However, towel and tissue mills cannot take all the available types of paper in mixed paper.

"We will take anything that comes across a desk, except food and boxes, and anything that comes in the mail, except boxes," says Nadine Mariconda, recycling coordinator for Marcal Paper Mill (Elmwood Park, NJ). "Newspapers are not a problem in offices as long as they are no more than 5% of the mixed paper. Brown envelopes from homes are not a problem because we get so few of them," adds Mariconda. Newspapers and cereal boxes are not wanted because their fiber is too weak. Because Marcal makes "100% recycled fiber" white tissue and towel products, it does not use chlorine bleach. "Browns" are a contaminant. Although "neon" colored paper and goldenrod pose deinking problems, they are found in limited quantities in offices, according to Mariconda. As a result, Marcal will allow them in the office paper mix in order to ensure high participation in the recycling program.

"We didn't want to use the word waste," says Jim Jenkin, fiber procurement coordinator for Kimberly-Clark (Dallas). "This is our raw material, not garbage. We can get good fiber from office buildings."

Kimberly-Clark buys two office paper grades, sorted office fiber and sorted daily office fiber. The sorted office grade has a 5% contamination level. The sorted daily office grade has a higher allowable contamination level (15%). Jenkin compares it to what normally comes out of an office building. Unbleachables, such as goldenrod-colored paper, "red-rope folders," chipboard and kraft (brown) envelopes are among the contaminants. Some materials are not acceptable at all. This includes obvious nonpaper products such as cans and bottles and less obvious products such as the envelopes made from high-density polyethylene fibers which are often used by overnight express-mail companies.

A paper mill making a brown paper hand towel for the commercial market can accept the unbleachables. The Fort Howard Corp. (Green Bay, WI) has three different grades for office paper. In addition to "file stock" and "office fiber pack grades" is the "ecosource" grade, which takes all office paper except restroom and cafeteria wastepaper. A Fort Howard subsidiary sorts the "ecosource" material to produce a raw material for specific products, with the browns destined for brown towels.

Finally, recycled paperboard (boxboard) mills will take the widest "mix" of waste paper. Boxes produced by these mills have different strength, brightness, and other needs than printing and writing paper or tissue paper.

Even these mills will not take all paper. While newspapers and browns are acceptable, waxed and polycoated paper- and food-contaminated paper can be excluded by a mill. "If you can rip it, we can take it," is how Murrel Smith, Jr., chief operation officer for Chesapeake Paperboard (Baltimore, MD) described his mill's specifications. Smith adds that cereal box liners are not paper. If it is not eliminated in the cleaning system, the wax on cereal box liners can cause boxes to be slippery and hurt the stackability of the boxes.

Mills differ. Even mills making the same type of product differ. An acceptable office pack for one mill may not be acceptable for another. Office packs will be more common for new mills using nonchlorine bleaches in their deinking. Colored paper can cause particular problems for nonchlorine bleaches, particularly the "high tone" colors such as neons and goldenrod. Ream wrappers from copier paper cause problems for some mills because they either use a plastic sheet or a moisture barrier to protect the copier paper. Mixed paper recycling programs must be aware of end market needs before they begin collecting paper.

Office Pack

The paperless office was supposed to be the wave of the future. Countless newspaper stories in the 1970s described how new information technologies would eliminate paper from business offices. In the 1990s reality turned out differently. The paperless office is now the urban forest, ready to be cut and made into new paper products.

Because they are so widespread in their scope and application, new office paper recycling programs could even be equated with clear-cutting.

Studies of office waste show just how paper-rich offices are. In 1990, offices generated 15.5 million tons of waste, according to a study by Franklin Associates for the National Office Paper Recycling Project (NOPRP, Washington, DC). Paper is 70% of this, or 10.9 million tons. Of this, 8.1 million tons are printing and writing paper. The remaining waste, glass, metals, plastics, trash, and food, constituted 30% or 4.6 million tons. NOPRP projections for 1995 show an additional 1.1 million tons of printing and writing paper in offices.

The Franklin study broke the paper fraction into printing and writing paper, newspaper, and corrugated boxes. Printing and writing paper consists of a wide variety of end uses ranging from business forms and copy paper to envelopes and books (see Table 5.1).

Clearly, the amount and relatively high quality of wastepaper available in office buildings makes these buildings a prime candidate for paper recycling programs. As the paper industry needs more recovered paper, office recycling programs are expanding what they collect and are making participation easier.

Recycall is an Atlanta company specializing in setting up office paper recycling programs. Recycall's success in high-grade programs, according to Bo Edwards, director of marketing for Recycall, is based

Table 5.1 Estimated Generation of Printing-Writing Papers in Offices, by End Use, 1990 (in thousands of tons)

Business forms	2,510
Reprographics (copy paper, ledger)	2,360
Commercial printing composite	1,750
File folders	480
Magazines	440
Stationery and tablets	260
Envelopes	230
Books	100
Other	110
Total	8,240

Source: Supply of and Recycling Demand for Office Waste Paper, 1990 to 1995, Franklin Associates, Prairie Village KS, 1991.

on stressing employee education, selecting the right containers for the work station and intermediate containers that fit in with the company esthetic, and strong management support. "I see a greater future for the office pack," he says, "it allows us to go at a greater portion of the market than before and eliminate many intermediate containers."

In Ohio, the Rumpke Co. (Cincinnati) is setting up office pack collection programs. Steve Sargent, director of recycling operations for Rumpke, says he is excited about the programs. "Our emphasis is on bulk collections using a packer truck. If a customer fits on a collection grid and can produce a minimum of two 90-gallon carts a week, we can offer this service," says Sargent. The truck can collect five or six tons with a minimum of compacting. According to Sargent, Rumpke produces an office pack which is a "hybrid file stock" that is not as clean as a white ledger grade, but is better than a mixed office grade.

As to value, don't expect to get paid for the office pack paper. "If the collection is efficient, we will take it for free, if we have to go into the building to collect the paper, we will charge for the service," says Barbara Drake, president, Covenant Recycling Services, Jacksonville, Florida. After the paper is collected, it must be processed into a raw material for an end-user.

If the paper has no value, why do employees in office buildings participate in office paper recycling programs? In some parts of the country, they have no choice; participation is mandatory as a result of state or local law. However, as one wastepaper dealer observed, "with a convenient, easy system, and an employee ethic that [it is] the right thing to do, people will participate."

Residential Mixed Paper

"I hate mixed paper. I've tried to sort it myself," is how one wastepaper processor reacts to the idea of residential mixed paper recycling. Collecting and processing residential mixed paper is harder and more complicated than office paper.

Where office paper is rich in strong, bleached fibers, residential paper is composed of a much wider mix of fibers, including a higher proportion of groundwood and unbleached fibers. As a result, collection and processing can be far more complicated than for office mixed paper (Table 5.2).

Table 5.2 Composition of Residential Mixed Paper in Curbside Recycling Program

Groundwood:	
Magazines	22.0
Newsprint	15.9
Catalog	2.4
Phone books	1.2
Other	0.2
Total	42.1
Corrugated boxes	16.3
Paper bags, sacks	4.8
Boxboard	8.9
Ledger paper:	
White ledger	8.0
Colored ledger	1.8
Other	0.2
Total	10.0
Junk mail	8.7
Other	9.2

NOTE: Newspaper was collected separately from residential mixed paper in the sampled area. Newspaper in this sample was included with residential mixed paper.

Source: Matrix Management Group, cited in Developing Markets for Recycling Multiple Grades of Paper, Environmental Defense Fund, New York City, 1992.

Understanding curbside collection of mixed paper is complicated by the fact that mixed paper is defined and collected differently throughout North America. In most places, mixed paper does not include cereal box liners, wax coated paper cartons, and food-contaminated paper. In some places, mixed paper is collected with newspaper and in some places it is collected separately.

In Seattle, the north-side collector uses a three-bin system with one bin used for mixed paper, one for newspaper and one for mixed bottles and cans. The south-side collector uses a 90-gallon cart in which all newspaper, mixed paper, cans, and plastic bottles are mixed together, and a separate container is used for glass bottles.

In Portland, homeowners sort 10 different materials and place them in separate bags into two yellow bins. Mixed paper is kept separate from magazines and newspapers. All materials are kept separate from each other in the collection truck. In Baltimore County, Maryland, residents place all mixed paper into a bag or tie it together with twine. Bottles and cans are placed in a blue bag. Haulers then collect the recyclables in a regular compactor truck, which is washed out so that the recyclables are not contaminated by garbage. In Quinte, Ontario, boxboard and household paper are placed in a separate sack, which is placed in a blue box.

Because of the variety in what is collected and how it is collected, curbside mixed paper programs report widely differing results. Portland is collecting 6.5 pounds per household of mixed paper per month from eligible households as opposed to 12 pounds per household in Seattle. In Quinte, Ontario, less than 30 pounds of boxboard per household, per year is being collected.

Collecting mixed paper has its own problems. "Scrap paper is light and fluffy, not heavy and dense, like old newspaper," according to Bruce Walker, manager of recycling programs for the city of Portland, Oregon. This creates a logistical problem as haulers have to unload the collection truck when the first compartment is full. Walker predicts that the addition of mixed paper could lead Portland to collecting recyclables in a more commingled fashion.

Perhaps, however, the biggest problem is that mixed paper generally has negative value. Some tissue mills will pay for residential mixed paper (Marcal will pay $5 per ton for loose paper). In Ontario, boxboard mills will pay $20 (Canadian) per metric ton for boxboard. Programs collecting a truly mixed grade must expect to pay a tipping fee to a processor. Baltimore County, Maryland, pays a $24.50 per ton tip fee to its processor.

Yet, despite a negative value, communities add mixed paper to their list of recyclables. "If the definition of a program is to maximize diversion from a landfill or incinerator, mixed paper is the way to go. If it is cash for trash, source separation of newspaper, magazines, or corrugated is the way," argues Chesapeake Paperboard's Smith.

A Mixed Future for Residential Mixed Paper?

"There is not enough pre-consumer and office source to meet potential demand by 1997. If you add the printing and writing paper from households, that will be enough to meet demand," says Ed Sparks, manager, recovered paper strategy and sourcing for Scott Paper (Philadelphia). The packaging market, in particular, will be losing traditional supplies of its paper to printing and writing and tissue mills. "Packaging mills will have to mine household paper to meet their needs," Sparks adds.

Nonetheless, residential mixed paper recycling has a long way to go. Making collection easy for residents and collectors while controlling processing costs is a major challenge.

"We should keep mixed paper separate from newsprint," argues Bob Davis, vice president, recycling systems for Browning-Ferris Industries (Houston). "In mixed programs, newspaper is half of what is collected and the cost to upgrade to a deink quality is horrendous."

When Halton, Ontario, recently added 5 items, including boxboard, to its curbside collection, it switched from weekly to every-other-week collection. Two-compartment collection trucks are used, with one compartment for mixed paper and the other for mixed containers. As a result, the pressure is on the processor to minimize its costs. Yet, boxboard markets are not paying enough to cover processing costs, according to Gwen Discepolo, CEO, Bronte 3Rs Material Recovery Facility (Oakville, Ontario).

Markets for mixed paper are further limited by geography. Many areas of the country, particularly the South, do not have ready access to paperboard mills. In addition, paperboard mills will limit the amount of residential paper they accept to ensure that they will have a clean enough furnish.

More importantly, adding mixed paper can lead to major changes in how recyclables are collected and will require additional education of householders. After all, when curbside programs add magazines to the list of recyclables, small amounts of magazines are collected, especially when magazine recycling starts. The recycled newspaper industry has been very successful at getting people not to recycle their magazines, and old habits die hard. Today, however, most recycled newspaper mills now want old magazines.

Other problems await mixed paper recycling. Towns tied into put or pay contracts with incinerators will probably not be able to start mixed paper programs. Towns with access to both tissue and boxboard mills will be in a dilemma. If they choose to supply the tissue mill, they will collect less paper and achieve a lower diversion rate than if they choose to supply the boxboard mill. They will also have a chance to get some cash for the paper. Finally, materials recovery facilities used to processing newspaper, bottles, and cans will find processing mixed paper far more challenging.

However, curbside recycling programs will collect more grades of paper over the next five years. Programs are already becoming "mixed paper" programs by adding magazines and corrugated boxes. Other grades will follow. The transition will not be easy and it will not be cheap, but it will happen.

6 SAFETY ISSUES

Although a shorter version of this article appeared in Waste Age, this longer, fuller version is being published here for the first time.

SAFETY TRAINING FOR WORKERS IN MATERIAL RECOVERY FACILITIES

By *John A. Legler* and *Richard B. Curtis*
Legler is executive vice president and Curtis is senior manager, equipment and safety, the Waste Equipment Technology Association, a sub-association of the Environmental Industry Associations, Washington, D.C.

The United States Occupational Safety and Health Administration (OSHA) has discovered MRFs! The safety agency has conducted an ever increasing number of on-site inspections in material recovery facilities over the last several years. As the recycling sector matures into new forms and more sophisticated methods of identifying, classifying, and sorting, it has tweaked the interest of the safety agency, which cut its teeth on regulating other material processing industries. As a result, there is a cry from the land for assistance in upgrading the safety performance of MRFs.

The National Solid Wastes Management Association (NSWMA), the predecessor to the Environmental Industry Associations (EIA), anticipated this in 1991 and launched its Processing Facilities Task Force to examine the problem. This group issued recommendations that the industry should endeavor to develop standards for various types of waste processing facilities, and resulting guidance documents and training programs to aid the recycling sector in complying with

OSHA mandates, reduce injuries, and mitigate losses which result from accidents.

Based upon the recommendations of the NSWMA Task Force the (American National Standards Institute) Accredited Standards Committee Z245 on Refuse Collection, Processing, and Disposal authorized a subcommittee (designated Z245.4) to develop draft standards. The Waste Equipment Technology Association has provided funding, technical expertise, and staffing for the Z245 Committee's activity since 1974. Because OSHA recognizes all of the Z245 Series as reference documents for enforcement activities, the eventual output of a MRF standard holds tremendous potential for the industry to have a key voice in how it is regulated.

While the Z245.4 subcommittee is still several years away from completion of a document for public comment, it has authorized EIA to pass recommendations along to the industry based upon its work in progress. While it must be cautioned that consensus has not been reached on these criteria under the rules of due process of the American National Standards Institute, facility operators may wish to avail themselves of this information in the development of their own safety and health programs (Table 6.1).

The section of the draft standard which is nearest to completion as of the writing of this narrative is that covering training requirements for workers. This chapter presents an overview of items which may be considered in the development of a formal educational curriculum.

The balance of this chapter presents the various criteria that have been identified as having a safety related component or affecting the safety and health of workers, contractors, facility users, or the general public. Facility designers, operators, manufacturers, and others involved in the planning, construction, operation, and maintenance of MRFs should use this information as a reference tool and "safety checklist."

While many of the concepts contained herein shall likely receive treatment in the eventual Z245.4 standard, the reader should be cautioned that this information, while comprehensive, is by no means all exclusive. The safety performance of a given facility is in many respects dependent upon the design of the individual facility, the types and flow of wastes, and the unique applications and combinations of technology that may be employed.

Table 6.1 Waste Collection and Processing Industrial Safety Standards

FIRE PROTECTION -

NFPA-82 - Fire Protection for Waste Compactors and Waste Storage Rooms.
NFPA-FPH, Section 12.7 - Fire Protection for Material Handling Equipment.
NFPA-101 - Life Safety Code.

SAFETY -

ANSI Z535 Series - Accident Prevention Signs and Marking of Physical Hazards.
ANSI Z4.1 - Sanitation in Places of Employment.
ANSI Z89 - Personal Protective Equipment-Headwear.
ANSI Z87 - PPE - Eye & Face Protection.
ANSI Z88.2 - PPE - Respiratory Protection.
ANSI Z49.1 - Safety in Welding & Cutting.
ANSI Z41 - PPE - Footwear.
ANSI Z244.1 - Lockout/Tagout of Energy Sources.
ANSI A1264 - Safety Requirements for Floor & Wall Openings, Railings, Tow Boards and Fixed Stairs.
ASTM E985-87 - Standard Specification for Permanent Metal Railings Systems and Rails for Buildings.
ASTM E849-86 - Standard Practice for Safety and Health Requirements Relating to Occupational Exposure to Asbestos.

ELECTRICAL SYSTEMS -

NFPA-70 - National Electrical Code.
NFPA-70E - Electrical Work Practices.

WORKER ENVIRONMENT -

ASTM E884 - Sampling Airborne Micro-Organisms in Solid Waste Processing Facilities.
ASTM E1076 - Health & Safety Record keeping at Solid Waste Processing Facilities.
ASTM E1132 - Occupational Exposure to Quartz Dust.
ASTM E1156 - Occupational Exposure to Silicates.

MOBILE EQUIPMENT -

ANSI/ASME B56.1 - Standards for Lift Trucks.
ANSI/ASME B56.7 - Standards for Crane Trucks.
ANSI/ASME B56.11.4 - Load Handling Symbols for Industrial Trucks.

Table 6.1 Waste Collection and Processing Industrial Safety Standards (Continued)

MOBILE EQUIPMENT (continued)-

ANSI/NSWMA Z245.1 - Safety Standards for Mobile Collection and Compaction Equipment.
SAE J994b - Backup Alarms.
SAE J1096 - Noise Measurement for Heavy Trucks.

STATIONARY PROCESSING EQUIPMENT -

ANSI/NSWMA Z245.2 - Safety Standards for Stationary Compactors.
ANSI/NSWMA Z245.5 - Safety Standards for Balers.
ANSI Z268.1 -Safety Requirements for Scrap Processing Equipment (Shears & Shredders).
ANSI/ASME B11.4 - Safety Requirements for Shears.
ANSI/ASME B11.1 - Safety Requirements for Mechanical Power Presses.
ANSI/ASME B11.2 - Safety Standards for Hydraulic Power Presses.
ANSI/ASME B15.1 - Guarding for Mechanical Power Transmission Apparatus.

CONTAINERS -

ANSI/NSWMA Z245.30 - Safety Requirements for Waste Containers.
ANSI/NSWMA Z245.6 - Rear Loader Container Compatibility Dimensions.

CONVEYORS -

ANSI/ASME B20.1 - Safety Standards for Conveyors.
ANSI/ASME E868 - Performance Tests on Resource Recovery Conveyors.
ANSI/CEMA - Series on Conveyor Applications.

CRANES -

ANSI/ASME B30 Series on Crane Safety.

Introduction

This chapter is intended to provide an overview of criteria which should be considered to address a very broad spectrum of possible hazards. Each waste processing facility designer and operator should evaluate the use of these suggested practices as they relate to actual situations encountered, and add information from individual experience and expertise, and other sources of information.

These practices are in no way meant to contradict any regulatory requirements, national consensus standards, or manufacturers' operating instructions. Those documents should be a part of all safety programs and should be treated as the primary guidelines.

The collection, processing, and transportation of wastes and recyclable materials is a highly specialized industry. Many factors affect the type of equipment and techniques chosen to handle the problems presented by recycling in a given area.

Waste processing equipment must be chosen which is durable and capable of containing, compacting, baling, separating, and moving various components of the waste stream. The truck weight, truck access, noise regulations of the individual's jurisdictions, as well as the routes to be traveled to and from the waste processing facility will often affect or limit the types of waste collection vehicles that may be used to supply materials to the facility. The mix of automated and manual techniques in residential collection is to a great extent dictated by terrain, street layout, and access to waste collection points.

No small part in the factors which govern the design of the waste processing facility and equipment is played by the regulators and planners which have jurisdiction over a particular waste shed. Likewise, the demands by markets for volumes and quality of various materials will have an effect on the economic viability and utility of the design of the waste processing facility. The factors are currently extremely volatile and uncertain as to future direction.

Thus, the hazards present in each operation are to a certain extent unique to the individual setting. It is therefore impossible to write a guidance document to cover all of the combinations and permutations of safety problems which can occur in the many potential designs of waste processing facilities. It is incumbent upon each waste processing facility risk manager or safety officer to evaluate the conditions present in his or her own operation. Operators who use various types of equipment and waste processing techniques may need to consider varying approaches to safety in different applications as well as a certain amount of flexibility to be able to adapt to changing conditions.

Safe operations are paramount to efficient operations. Incidents and accidents reduce time available for productive work and ultimately affect bottom line economic performance. The employee is the key to safe operations. Each vehicle or machine operator, sorter, and supervisor must be thoroughly knowledgeable about the equipment and specific job task as well as the types of materials to be handled. Safety is to be placed in a priority position over all other factors. Operators of equipment that is capable of handling, crushing, baling, and compacting large quantities of material have a great responsibility to ensure their own safety as well as that of their fellow employees and the general public.

The employer is the critical link in the safety chain, as the bridge between product producers, system designers, safety and environmental regulators, and the ultimate end-user (the line employee). Waste processing facility operators must ensure that all available information is gathered and is transmitted to those responsible for their own safety and that of others. Effective management of a formal safety program and ongoing training/communication programs helps fulfill the employer's charge under the Occupational Safety and Health Act to provide a safe working environment.

Training

General

The Occupational Safety and Health Act, under which the federal OSHA derives its regulatory authority, requires that employers maintain a workplace free from known hazards which create the potential for injury or illness. Workers are also required under the Act to adhere to the safety policies and procedures set by their employers. Organized training is the means through which workers learn their duties and responsibilities, the hazards that they may encounter in fulfilling their duties, and the means by which they can protect themselves, their fellow workers, and others, such as facility users, contractor personnel, and the general public.

Facility operators are responsible for training all employees, including supervisors, appropriate to their assigned jobs and tasks. While the primary obligation for independent contractors and labor pool providers is to train their own employees, a facility operator has a responsibility as well to ensure that known hazards are communicated to contractors.

In many cases, the operating instructions provided by the original manufacturers of the equipment used in a facility must be augmented by the facility operator in order to be properly protective of the workers in the actual application. Consideration must be given not only to the operation of the equipment as part of the processing line of which it is part, but also of other workers' activities that are affected due to their proximity to the system. The employee who is authorized to operate particular machines and systems certainly must be trained in the functioning of that technology, but affected workers have a need to know at least the basic functions of machines near their work stations, how to recognize hazardous situations, and how to respond in the event of an emergency, as well as what not to do in order to avoid creating an unanticipated hazard.

Training provided to contractors should be appropriate to the jobs and tasks that may be performed by the contractor within the facility. In most cases, contract laborers should be treated the same as regular facility employees. Adequate training should be provided either by the contract laborer's parent employer or by the facility operator where a job or task is performed.

Training of independent contractors is a nettlesome issue, but consider that untrained contractors working next to regular MRF workers may, by their lack of having been taught proper safety procedures, present a hazard in and of themselves. This places at least a secondary responsibility upon the facility operator for all individuals performing duties in the facility.

Training Needs To Be Organized and Periodic

Training should be provided for all employees at least upon initial assignment to a job or task, with periodic refresher training as necessary to maintain the required level of competence. Retraining needs to be considered for employees whenever there is a change in their job assignments, or a change in machines, equipment, or processes that may present a new hazard. Additional retraining should also be conducted whenever a periodic inspection and supervision reveals deviations from procedures by workers or inadequacies in an employee's knowledge or use of procedures.

OSHA's own research shows that when this information is utilized in an organized written program, safe work practices are readily understood and followed by employees, and accidents are significantly reduced. In studies leading up to the government's standard for the control of hazardous energy sources, OSHA found that workers who had only received "On the Job Training" were involved in accidents 3.5 times more often than workers who had access to formal, written training programs. In promulgating the regulation on control of hazardous energy sources, OSHA cited statistics of the Bureau of Labor. Some valuable lessons can also be derived from looking at the training levels of injured workers in the BLS report:

Underlying Training of Injured Employees:

Received printed instructions	4%
Procedures posted on equipment	6%
On-the-job-training or meetings	34%
No training at all	55%

Over half had received no training whatsoever. In effect, they were experimenting with their well-being when performing equipment service. Another third received training by the means most often used in general industry; that is, through "on the job" training. Where employees, in addition to training, could read instructions regarding energy control posted on the equipment, far fewer accidents occurred. OSHA cited verbal instruction as a barrier to effective communication in this manner. When employees are given printed instructions to which they can refer during home study or during servicing operations, very few accidents occur.

The employer should maintain records of training to include the date(s) of the training and the content of training received. Records need to be maintained for a minimum of three years, unless otherwise specified by a specific OSHA rule. Recordkeeping deficiencies are among the most frequently cited violations by OSHA, as the agency considers logging of education to be a key part of safety management controls, and an important indicator of the employer's practices and attitudes toward safety.

Where To Find Information

The primary guidelines are OSHA's regulations for General Industry, 49 CFR Part 1910 (available from the federal Government Printing Office). OSHA's rules provide the basic framework for all training, but are performance based standards. In other words, they tell you the subjects to teach, but not the substance. Other sources need to be consulted in order for a facility operator to customize its program to its own circumstances. Most employers would be able to find sufficient resources from a combination of the rules themselves in the references listed in Table 6.2.

Employers need to look closely at the manufacturer's, installer's, modifier's or system designer's instructions, operating and maintenance procedures, and work practices. In addition, industry standards produced by accredited developers such as the Z245 Committee (available through EIA), other ANSI accredited committees, and the

Table 6.2 Regulatory OSHA Standards

PART 1904 - RECORD KEEPING

PART 1910 - SAFETY & HEALTH STANDARDS

- Subpart C - Access to Employee Medical Records
- Subpart D - Walking/Working Surfaces
    - Stairs & Ladders
    - Floor & Wall Openings
    - Scaffolds (Proposed changes include exposure to falls)
- Subpart E - Means of Entry
- Subpart F - Manlifts & Platforms
- Subpart G - Environmental Control
    - Ventilation
    - Noise
    - Non-ionizing Radiation
- Subpart H - Hazardous Materials
    - Acetylene
    - Flammable & Combustible Liquids
    - LPG
    - Spray Finishing (Painting)
- Subpart I - Personal Protection Equipment
    - Eye & Face Protection
    - Respiratory Protection
    - Head Protection
    - Foot Protection
    - Electrical PPE
- Subpart J - General Environment
    - Signs & Tags
    - Sanitation
    - Hazard Markings
    - Control of Hazardous Energy Sources (Lockout/Tagout)
    - Confined Space Entry
- Subpart K - Medical & First Aid
- Subpart L - Fire Protection
    - Fire Brigades
    - Fire Suppression Equipment
    - Fire Protection Systems
- Subpart M - Compressed Air Systems
- Subpart N - Materials Handling & Storage
    - General Rules
    - Powered Industrial Trucks
    - Cranes
- Subpart O - Machines & Guarding
    - General Rules
    - Power Presses
    - Mechanical Transmission Apparatus

Table 6.2 Regulatory OSHA Standards (Continued)

- Subpart P - Portable Power Equipment
- Subpart Q - Welding, Cutting & Brazing
- Subpart R - Special Industries - Paper Mills
- Subpart S - Electrical
  - Systems
  - Work Practices
  - Maintenance
- Subpart Z - Hazardous Substances
  - Air Contaminates
  - Asbestos
  - Bloodborne Pathogens
  - Hazard Communication Standard

OSHA PROPOSED RULES NOW IN FINAL RULEMAKING STAGE

- Revised Personal Protective Equipment
- Revised Walking/Working Surfaces
- Vehicle Occupant Protection

OSHA EXPECTED PROPOSALS FOR NEW RULES IN THE 1990s

- Ergonomics & Cumulative Trauma Disorders
- Workplace Hazard & Medical Monitoring

National Fire Protection Association should be part of everyone's safety library (contact the American National Standards Institute, 11 W. 42nd Street, New York, NY 10036, 212-642-4900).

Training products and guidance documents to assist in training are also available from many sources. NSWMA will be preparing a safety guidance document for MRFs in parallel to the development of the Z245.4 project, its Manual of Recommended Safety Practices provides a primer in a number of areas, and provides periodic advice to the industry through articles in *Waste Age* magazine. Two excellent libraries of OSHA compliance training materials include the National Safety Council, P.O. Box 11933, Chicago, IL 800-621-7619 and J.J. Keller & Associates, 145 W. Wisconsin Ave., Neenah, WI 54957, 800-242-6469.

Training Curriculum for MRF Workers

A suggested set of training requirements is outlined in Table 6.3. These requirements are in accordance with federal OSHA and other federal agency requirements, where they exist, and provide guidance where there are no specific regulatory standard obligations. In the event of ambiguity or conflict, federal standards always prevail.

EIA's recommendations are very performance oriented. In other words, the complexity and depth of the training provided must be geared to the facility in question. A simple, primarily manual MRF may be adequately served by a cursory overview of most of the subject matter. A multi-line, highly machine driven, mixed waste facility may present complex hazards and interactions among various sources of hazards which dictate in-depth classroom instruction before allowing workers to take on certain jobs or functions.

The curriculum is organized according to various jobs classifications typical in a MRF, and the safety topics which may need to be covered in a typical commingled facility with a moderate degree of mechanical technology. The training concept embodies basic training in the appropriate subjects (marked by "X" in the table. Additional detailed training as applicable according to tasks which present higher levels of hazard (marked "AA" in the table), creating a class of "authorized employees" whose specific training enables them to safely perform such jobs as operating balers, loaders, and forklifts, or perform as traffic spotters.

Employees whose work is near operations which require authorized employees need to know, for example, the meaning of a lockout/tagout situation involving a baler or process line machine. Such training should inform affected employees (marked "AFF" in the table) of the potential hazards which may occur and how they are to protect themselves through procedures, protective equipment, etc. Other employees may need to be given training to recognize certain hazards (marked as "REC" in the table), such as potentially infectious needles, so as to be able to avoid potential hazards or to notify emergency response personnel to clean up a spill.

Facility operators/employers are responsible for ensuring all employees, including supervisors, contractors, and contract laborers are properly trained appropriate to their assigned jobs and tasks. Training provided to contractors will be appropriate to the jobs/tasks that may be performed by the contractor. Training for contract laborers should be provided the same as for regular facility employees. The employer/facility operator is responsible for ensuring that adequate training is provided either by the contract laborer's parent employer or by the facility operator where a job/task is performed. Employers must observe state and local requirements where local authority has jurisdiction.

Table 6.3 MRF Training Curriculum

Jobs	Basic Hazcom	Walking Working Surface	Spill Resp	Bloodborne	Energy Cont	Confined Space	Ergo	Heat/ Cold	PPE	Hearing Cons	Traffic Cont	Process Equip	Powered Industrial Truck	Elec. Safety Prac.	Fire Safety	Specialty Equip	L/M SC	Respirator	Reports & Records	Substance Abuse
Scalehouse Operator	X	AA	Rec			AA	X		Basic	AA	X				X		IF		X	X
Sorter	X	X	Rec	Rec	Aff	Aff	X	Not Gen In AA	Basic	X	X	AA				AA	IF	AA		X
Spotter	X	X	Rec	Rec	Aff	Aff	X	X	Basic	X	X	AA	AA		X	AA	IF	AA	X	X
Powered Ind. Equipment Operators	X	X	Rec	Rec	X	Aff	X	X	Basic	X	X	X	X		X	AA	IF	AA	X	X
Baler/Comp/Proc. Op	X	X	Rec	Rec	X	X	X	X	Basic AA	X	AA		AA	X	X	X	IF	AA	X	X
Supervisor	X	X	X	X	X	X	X	X	X	X	X	X	X	X	X	X	IF	X	X	X
Maintenance/Service	X	X	Rec	Rec	X	X	X	X	X	X	AA	X	AA	X	X	X	IF	X	X	X
Housekeeping	X	X	Rec	Rec	Rec	Rec	X	X	X	X		X				AA	IF			X
Drivers	X	X	Rec	Rec	X	X	X		X	X	X				X	X	IF		X	X
Fire Brigades	X	X	X	Rec	X	X	AA	X	X		X	X		X	X	AA	IF	X	X	X
Emergency Response (All)	X	X	Rec	X	Aff	Aff	AA	X	X	AA	X	X		AA	X	AA	IF	AA	X	X
Confined Space Emergency Team	X	X	Rec	Rec	X	X	AA	X	X	AA	X	X		AA	X	AA	IF	AA	X	X
Spill Response Team	X	X	X	X	X	X	AA	X	X	AA	X	X		AA	X	AA	IF	AA	X	X
Hazardous Waste Team	X	X	X	Rec	Aff`	X	AA	X	X	AA	X			AA	X	AA	IF	AA	X	X
Clerical who enter floor	X	AA															IF			X
Administrative who enter floor	X	X	Rec	Rec	Rec	Rec			Basic		AA				X		IF			X
Contractors	X	X	Rec	Rec	Aff/AA	Aff/AA			Basic	AA	AA	AA		AA	X	X		AA	AA	X
Labor/Management Safety Committee	X	X	X	X	X	X	X	X	X	X	X	X	X	X	X	X	X	X	X	X

AA = As appropriate to specific job in addition & basic concept training. AFF = Affected Party Training. REC = "Recognition" Training Only.

Training should be provided at least upon initial assignment to a job or task, with periodic refresher training as necessary to maintain the required level of competence. Retraining should be provided for employees whenever there is a change in their job assignments, or a change in machines, equipment, or processes that presents a new hazard. Additional retraining should also be conducted whenever a periodic inspection reveals, or whenever the employer has reason to believe, that there are deviations from or inadequacies in the employee's knowledge or use of procedures.

Employers should refer employees to the manufacturer's, installer's, modifier's or system designer's instructions to ensure that correct operating and maintenance procedures and work practices are understood and followed.

The employer should maintain records of training to include the date(s) of the training and the content of training received. Records should be maintained for a minimum of three years, unless otherwise specified in a regulation or industry standard.

Training curricula requirements are outlined in Table 6.3. These requirements are in accordance with federal and state OSHA and other federal agency requirements. In the event of ambiguity or conflict, federal or state standards prevail. The topical training elements which should be considered include the following:

1. Site Safety Orientation
2. Basic Hazard Communications (HAZCOM)
3. Walking - Working Surfaces
4. Emergency Spill Response
5. Bloodborne Pathogens
6. Energy Control (Lockout/Tagout)
7. Confined Space Entry
8. Ergonomics
9. Heat/Cold Stress
10. Personal Protective Equipment (PPE)
11. Hearing Conservation
12. Traffic Control
13. Powered Industrial Trucks
14. Electrical Safety Practice
15. Fire Safety
16. Material Control

1. Site Safety Orientation

Employers should provide general site safety orientation and training to all employees and any personnel who are directly or indirectly involved with MRF operations to enhance personnel safety and health.

Training elements should include:

a. General work rules and regulations
b. General safety and health policy
c. Facility and processing equipment familiarization
d. Signs, accident prevention warnings, cautions and alarms
e. Emergency action
  1. Personnel
  2. Alarms, egress/evacuation, fire exits
  3. Rescue and medical duties, access to health care professionals, basic first aid, eye wash, etc.
  4. Procedures for fire, toxic chemicals, tornado, hurricane and other natural disasters
f. Accident reporting
g. Substance abuse
h. Discipline policy
i. The employer's training policy

2. Basic Hazard Communications

Employers are required to provide information to their employees about the hazardous chemicals or hazardous materials to which they are exposed by means of a HAZCOM Program, labels, and other forms of warning, material safety data sheets (MSDS), and information and training.

Training elements should include:

a. An explanation of the HAZCOM standard requirements and how the program applies in the workplace
b. How to read and interpret information on labels and material safety data sheets (MSDSs)
c. The location of the HAZCOM Plan and how employees can locate and use the available information
d. The physical and health hazards of the chemicals and hazardous materials in the employees' work areas

e. Protection measures against the hazards
f. Facility/company procedures to provide protection, e.g., safe work practices, emergency procedures, and the use of PPE
g. Methods to detect the presence of a hazardous chemical or material to which they may be exposed

3. Walking - Working Surfaces

Employers should provide training for employees in MRF walking and working surfaces. The program should instruct authorized employees to recognize and avoid the hazards associated with the areas in which they work. The employer should require that others, e.g., contractors, whose employees use the pit area provide assurance of training to recognize and avoid the hazards associated with the areas of the facility in which they perform their duties (Contractor access to the facility should be limited to only those areas necessary for the tasks to be performed).

Training elements should include:

a. Safety procedures for facility areas such as:
   1. Guard rails, covers, ladders, stairs, walkways, platforms, lane markings, aisles and passageways
   2. Floor loading protection
   3. Floor and wall openings and holes
b. General housekeeping procedures

4. Spill Response

Employers should develop a training program for all employees exposed to safety and health hazards during hazardous waste operations. Supervisors and workers must be trained to recognize hazards and to prevent them. The amount of instruction differs with the nature of the work operations, with each employee being trained to the level required by their job function and responsibility. Under normal circumstances emergency spill responders will be in the category of "first responders" who require minimum awareness training. These individuals are likely to witness a spill in the release area and should be trained how to notify proper authorities or respond only in a defensive fashion without actually trying to stop or clean up a spill. They should be trained to contain the release from a safe distance, keep it from spreading, and prevent exposure.

In circumstances where employees are more likely to encounter hazardous waste materials at the MRF, employers should provide comprehensive training, required by federal, state, and local agencies, to employees in emergency response cleanup operations involving hazardous substances. An emergency response is one required by employees or designated responders outside the immediate release area. This does not include a response to incidental releases which can be controlled by employees in the release area, or to releases of substances where there is no safety or health hazard.

Training elements should include:

a. Employer Safety and Health Program
b. Emergency Response Plan
c. Hazard identification
d. Emergency alert and reporting procedures
e. Personnel responsibilities
f. Evacuation/places of refuge
g. PPE
h. Decontamination/washing facilities
i. Medical treatment/first aid
j. Engineering controls
k. Work practices/procedures
l. Contractor/subcontractor requirements

5. Bloodborne Pathogens

Bloodborne Pathogen hazard training should be provided to the employee prior to initial assignment to a job, and at least annually thereafter. Training records should be retained for at least three years. Employers, as a minimum, should provide awareness training to employees regarding:

a. The hazards that may be encountered from exposure to bloodborne pathogens, including terminology and definitions; communication of the hazards/risks, including the possibility of being infected with Hepatitis B or HIV
b. Training on the proper incident reporting procedures, first aid and/or medical services that are available, and spill response procedures to follow in the unlikely event of an exposure incident

If occupational exposure to bloodborne pathogens is reasonably anticipated, employers should provide awareness training to employees and, in addition, comprehensive training in accordance with 29 CFR 1910.1030, *Bloodborne Pathogens*, to include:

c. Employer's Exposure Control Plan
d. Availability of vaccines
e. Personal protective equipment
f. Procedures for exposure treatment and follow-up

6. Energy Control (Lockout/Tagout)

Employers should provide training to affected employees to ensure that the purpose and function of the energy control program are understood by the employees and that the appropriate knowledge and skills regarding energy controls are acquired by the employees where unexpected energizing or starting up of the machine or equipment or release of stored energy could cause injury.

Training elements should include:

a. Recognition of hazardous energy sources
b. The type and magnitude of the energy available in the workplace
c. Energy control program
   1. Lockout/tagout procedure/sequence
   2. Limitation of tags
d. Procedures for restoring equipment to service

Initial training should be provided to all affected employees as well as all employees whose work operations are or may be in an area where energy control procedures may be utilized. Retraining should occur whenever there is a job change or a change in machines which presents a new hazard, or when there is a change in the energy control procedures.

7. Confined Space

Employers should provide training to employees to ensure that employees are protected from the hazards of entry into permit-required confined spaces (permit spaces) found in the workplace.

Training elements should include:

a. Definitions/classification of spaces
   1. Hazardous atmosphere
   2. Confined space
   3. Permit-required space
b. Risks/hazards
c. Permit-space program/procedures (if required)
   1. Isolation, purging, and hazard control
   2. Permit procedures
   3. Equipment (PPE)
   4. Atmospheric testing
   5. Employee responsibilities
   6. Contractor responsibilities
d. Rescue and emergency procedures

All employees who may encounter permit-spaces in the workplace should receive training before initial assignment to permit-space duties, before there is a change in assigned duties, and whenever there is a change presenting new hazards. Training should include annual drills in simulated rescues, basic first aid and CPR for rescue teams.

8. Ergonomics

Employers should provide training to ensure that employees are sufficiently informed about ergonomic hazards to which they may be exposed so that they may be able to participate in their own protection. Training should be provided to all affected employees and their immediate supervisors.

Training elements should include:

a. Employer's ergonomics/medical management program
b. Awareness of potential risks, hazards
c. Varieties of cumulative trauma disorders (CTDs)
d. Recognizing and reporting causes, symptoms
e. Prevention and treatment
   1. Tools: care, use and handling techniques
   2. Guards and safety equipment
   3. Body mechanics, lifting techniques and devices
   4. Work methods/procedures
   5. Work station design
   6. Job/task rotation

9. Heat/Cold Stress

Employers should provide training in order to prevent or reduce the risk of adverse safety and health effects to employees exposed to workplace heat and cold stress.

Training elements should include:

a. Heat and cold stress awareness, recognition, and causes
b. Definitions
   1. Heat stress
   2. Cold stress
c. Facility risks and hazards
d. Recognition, signs, symptoms
e. Engineering or administrative controls to reduce/control stressful conditions
f. Prevention and treatment procedures
   1. Work and hygienic practices
   2. Remediation techniques
   3. PPE/clothing
   4. Risk reduction
   5. Basic first aid

10. Personal Protective Equipment (PPE)

Employers should provide training to employees for personal protective equipment (PPE) for eyes, face, head, and extremities, protective clothing, respiratory devices (if required), and protective shields and barriers which should be employed for protection against hazards of processes or the environment, chemical hazards, radiological hazards, or mechanical irritants which may cause injury or impairment to the employee in the performance of his/her work.

Training elements should include:

a. PPE requirements
   1. Reasonable probability of need to use
   2. Hazards, risks
b. Types of equipment
   1. Limitations and precautions
   2. Equipment availability
c. Policies and operating procedures
   1. Equipment storage, inspection, care and maintenance

2. Equipment selection and use (see Table 6.3)
3. Respirator (if required) fitting, demonstration, and practice

11. Hearing Conservation

Employers should provide awareness training to employees regarding the hazards/risks that may be encountered from exposure to high noise levels and basic hearing conservation measures. For all employees who are exposed to noise levels at or above an 8-hour time-weighted average of 85 decibels, the employer must, in addition to awareness training, institute a training program, and ensure employee participation in the program.

Training elements are to include:

a. The effects of noise on hearing
b. Noise levels and exposure limits
c. Hazards/risks specific to work environment
d. Protection/prevention
   1. The purpose of hearing protection
   2. Advantages, disadvantages, and attenuation of various types of protection
   3. Instruction on selection, fitting, use, and care of protective equipment
   4. Requirements for and purpose of audiometric testing, and an explanation of the test procedures
   5. Availability of information and training on hearing conservation
   6. Applicable state/local regulations

12. Traffic Control

Employers should train their employees in MRF operational procedures to ensure that employees are aware of the hazards of vehicular traffic in and around the facility.

Training elements should include:

a. General MRF
   1. Traffic
   2. Traffic signals/markings

   3. Public/commercial traffic routes
   4. Hazards, common mishaps/accidents, unsafe practices
   5. Accident reporting
b. Vehicle types
   1. External/contractors
   2. Internal/powered industrial trucks.
c. Pedestrian routes/walkways
d. Railway traffic (if applicable)
e. Safety/Rules and regulations
   1. General
   2. Employee visibility, including requirements for wearing enhanced visibility clothing

13. Waste Processing Equipment

Employers should provide training to all persons engaged in the operation, cleaning, maintenance, service, or repair of waste processing machinery and equipment contained in a material recovery facility to ensure that all persons are thoroughly familiar with and competent to operate and maintain the processing equipment to minimize the possibility of personal injury.

Training should be tailored for each facility varying from general awareness training provided for all affected employees to more detailed, equipment-specific training for equipment operators and maintenance personnel. Operator and maintenance training should include practical demonstration of equipment operation knowledge and skills by the employee. Employers are responsible for providing training as well as for properly authorizing employees to operate processing equipment.

Training elements should include:

a. Equipment knowledge
   1. Equipment
   2. Operating, safety, and emergency controls
   3. Equipment and machinery point-of-operation safeguards
b. Operator qualifications
   1. Training requirements
   2. Demonstrated skills
   3. Demonstrated knowledge
c. Operating procedures
   1. Operating limitations and restrictions

2. Operating instructions and manuals, including manufacturer's recommended practices and procedures
3. Equipment preoperational checks
4. Safety precautions, hazards, cautions, warnings, and alarms for applicable mechanical, pneumatic, hydraulic, and electrical components and systems; and, personal protective equipment

d. Equipment servicing, inspections and cleaning
1. Energy control (lockout/tagout)
2. Authorized procedures

e. Maintenance
1. Basic equipment troubleshooting
2. Regular, periodic and preventive maintenance requirements
3. Maintenance manuals, reports, and instructions
4. Manufacturer's requirements and recommended procedures
5. Maintenance, malfunction, and repair reporting and recordkeeping
6. Equipment modification/alteration policies and procedures.

f. Systems integration/interdependence

g. Systems safety

14. Powered Industrial Truck Operations

Employers must provide employees with the knowledge and skills to safely operate, service, maintain, and repair fork trucks, tractors, platform lift trucks, motorized hand trucks, and other specialized industrial trucks powered by electrical motors or internal combustion engines. Employers are responsible for employee training as well as for properly authorizing employees to operate equipment.

Training elements should include:

a. Equipment knowledge
1. Equipment description/nomenclature/familiarization
2. Load capacity
3. Safety guards

b. Operator qualifications
1. Training requirements
2. Test, authorization, and certification requirements

c. Operating procedures
1. Operating limitations and restrictions

2. Operating instructions and manuals, including manufacturer's recommended practices and procedures
3. Preoperational checks
4. Operating in hazardous atmospheres
5. Safety precautions, hazards, cautions, warnings, and alarms for applicable mechanical, pneumatic, hydraulic, and electrical components and systems
6. Use of brakes and wheel chocks
7. Control of noxious gases and fumes (i.e., exhausts)
8. Docking and loading
9. Pedestrian traffic
10. Material handling (e.g., bale and load stacking, etc.)

d. Traffic regulations and traveling procedures
   1. Situational awareness
   2. Speed limits
   3. Rights-of-way
   4. Horns, lights, and signals
   5. Maneuvering and loading

e. Servicing, inspections, and cleaning
   1. Fuel handling and storage
   2. Changing and charging storage batteries

f. Maintenance
   1. Basic troubleshooting
   2. Regular, periodic, and preventive maintenance requirements
   3. Maintenance manuals, records, and instructions
   4. Manufacturer's requirements and recommended procedures
   5. Equipment modification/alteration policies and procedures

15. Electrical Safety Practice

Employers must provide training on electrical safety requirements that are necessary for the practical safeguarding of employees who face the risk of electrical shock that is not reduced to safe levels or other injury resulting from direct or indirect electrical contacts in the workplace. Training should include electrical safety-related work practices for both qualified (those who have training in avoiding the electrical hazards of working on or near exposed energized parts) and unqualified persons (those with little or no training). The degree of training should be determined by the employer based on the job/occupational category and the risk to the employee.

Awareness training elements for affected employees should provide for knowledge of general hazards to include:

a. Location, knowledge of premises (including overhead lines), vehicular and railway (if applicable) wiring, conductors and equipment
b. Skills and techniques required to distinguish exposed live parts from other parts of electrical equipment
c. Skills and techniques required to determine nominal voltage of exposed parts

In addition to awareness training, authorized employees should be provided specific training, as appropriate, on job/task specific work practices to include the following elements:

a. De-energizing procedures
b. Control devices
c. Working on or near de-energized exposed parts
d. Lockout/tagout procedures
e. Stored energy
f. Re-energizing equipment
g. Working on energized equipment
h. Insulation
i. Use of flammable or ignitable materials
j. Safeguards, tools, and personal protective equipment
k. Vehicle operations in the vicinity of exposed electrical equipment or overhead lines
l. Working with ladders
m. Working in confined spaces
n. Conductive material and apparel
o. Cleanliness and housekeeping
p. Portable electrical equipment inspection, handling, and use
q. Safety signs and tags
r. Barricades and attendants
s. Manufacturer's guidelines (equipment specific)
t. Electrical system interdependence and system safety

16. Fire Safety

Employers should provide fire safety awareness training to employees as appropriate and provide additional task-specific training for incipient fire responders commensurate with those duties, functions, and responsibilities that the employee is expected to perform. Individual responsibilities should be delineated in the employer's organizational statement or operating policies.

NOTE - Employers normally are not required to maintain and train *interior structural fire brigade* members that require a higher level of training because of the increased hazards and risks involved.

Incipient fire responders must only combat a fire which is in the initial or beginning stage and which can be controlled or extinguished by a portable fire extinguisher, Class II standpipe, or small hose systems without the need for protective clothing or breathing apparatus. Incipient fire responders are not to combat interior structural fires or perform rescues inside of buildings or enclosed structures which are involved in a fire situation beyond the incipient stage.

Incipient stage fire responders are *NOT* to enter smoke-filled or toxic-filled environments where protective clothing or breathing apparatus are required. Moreover, search and rescue operations are to be conducted only by personnel trained in emergency operations such as members of an interior structural fire brigade, or an equivalent unit.

Incipient fire responders may be cross-trained as permit-required confined space entry team members. Training should be provided before specific duties are performed with refresher training conducted at least annually.

Awareness training elements for all affected employees should include:

a. Emergency action procedures
   1. Evacuation plans, exits, emergency escape routes (particularly designated handicapped routes)
   2. Personnel responsibilities
b. Fire reporting
   1. Alarm location and operation
   2. Personnel notification
c. Fire prevention procedures
   1. Housekeeping practices
   2. Material storage procedures
d. Hazards and special hazards in the workplace
   1. Location, use of flammable liquids, paint, gases, toxic chemicals, and water reactive substances
   2. Welding, cutting and brazing (if applicable)
      - Employee responsibilities
      - Special precautions, guards, restrictions, and prohibited areas
      - Combustible materials
      - Fire watch requirements

Incipient fire responders should receive awareness training, and in addition, training in the following topics:

a. Location and use of fixed and portable fire extinguishers, standpipes, sprinklers, and other fire equipment
b. Protective clothing and equipment requirements
c. Equipment inspection, maintenance and testing
d. Basic first aid medical procedures

17. Material Control

During the course of operating a facility, employees are required to handle and store a wide variety of materials. Employers are responsible for providing training to minimize the risk of employee injury and provide for the safety and health of their employees involved in material control functions.

Training should be in accordance with 29 CFR 1910 Subpart N, and should include:

a. Plant operations overview/familiarization
    1. Type of materials handled
    2. Plant physical layout
b. Facility policies/procedures
    1. Traffic routes/traffic flow
    2. Storage areas/shelving
    3. Housekeeping
c. Risks/hazards
    1. General safety precautions
    2. Fire
    3. Personal injury
    4. Stacking/tiers
    5. Load stability
    6. Floor loading
    7. Clearance limits
d. Mechanized equipment operation and use
    1. Powered industrial trucks
    2. Hoists, hand trucks, conveyors, chutes, hooks and clamps, tiedowns, wire ropes, and block and tackle rigging
    3. Loading and unloading transport vehicles
e. Engineering controls and work practices
    1. Ergonomics
    2. Lifting techniques
    3. Overexertion/stress

Training Holds the Key

In a number of states, led by California and Minnesota, employers are required to compile a formal written comprehensive safety and health plan. Given the myriad of potential safety concerns possible in MRFs, since they are all different, the construction of a plan can seem a daunting exercise even for the seasoned safety professional.

Training is the threshold element which unlocks the door for program formulation. In order to train, one must research issues and operations. In developing classroom materials, a trainer must have policies and procedures to reference and operations manuals in order to conduct demonstrations. As a MRF operator develops a training program, element by element, the comprehensive safety program will take place.

7 MANAGEMENT/ STRATEGY ISSUES

(How do you finance a recycling facility? And how does recycling fit into a solid waste management plan? To a large extent, most, if not all, of the articles in this book discuss the latter, seeing recycling as a component of a larger strategy, but a particular aspect of the discussion needs more elaboration. Waste-to-energy or municipal waste combustion has often been seen as the antithesis of recycling—one reuses material while the other turns it to ash and energy, but the second article in this chapter suggests that the antithesis argument lacks foundation. Similarly, source reduction is usually the first component in a solid waste management plan—and the least discussed. The final article in this chapter discusses it.

Finally, the issue of scavengers as a collection mechanism and a diversion of recyclables—a pertinent U.S. issue, given the number of homeless people and an international/cultural/way of life issue—is discussed.)

FINANCING RECYCLING PROGRAMS

By *Robert W. Ollis, Jr.*, a partner with the law firm of Chapman and Cutler in Chicago

Communities are now using a variety of methods to collect recyclable materials, including residential curbside collection, drop-off centers, buyback centers, and commercial recycling services. Such techniques often require large numbers of individual recycling bins, specialized trucks, transfer stations, and collection centers.

Once collection occurs, equipment is needed to separate mixed recyclable items or to remove valuable materials from a mixed waste stream before the materials can be sold. These materials recovery facilities (MRFs), intermediate processing centers (IPCs), or materials resource recovery facilities (MRRFs) require financial resources and technical expertise.

One way that local governments can provide the financing for the development of recycling facilities is to issue tax-exempt bonds, since such items typically qualify as "solid waste disposal facilities" under the Internal Revenue Code. Tax-exempt financing generally affords the lowest-cost form of borrowing for a large project, by either reducing the community's own expenditures or providing an incentive to a private company that intends to build or operate a recycling facility.

Public-Private Partnerships' Role

Many communities lack the financial resources or technical expertise to construct or operate the facilities that are needed for full-scale recycling programs. In order to promote recycling operations on a private basis, many municipalities have entered into contractual relationships with private companies—referred to as "public-private partnerships"—to collect recyclable materials or to own or operate the recycling facilities.

Tax-exempt financing is available both for government-owned projects as well as for those facilities involving private ownership or operation. The extent of a private company's involvement does, however, affect the type of tax-exempt bonds that may be used.

Two major types of bonds bear interest that is exempt from federal income taxation, "governmental use bonds," and "private activity bonds." "Governmental use bonds" are the most advantageous, and are available for those projects that will be owned and operated by a governmental unit. Such bonds typically provide the lowest tax-exempt interest rates and offer greater flexibility to a municipality. Governmental use bonds, however, have strict limits on the amount of private participation in the project.

In one Illinois example, a group of municipalities have formed a joint-action agency to finance a system of transfer stations, recycling facilities, and a large landfill. The agency wants to retain ownership of the project, but prefers to bring in a private company to run the facilities on a limited basis.

The agency should still be permitted to issue governmental use bonds, as long as they comply with the "management contract rules" of the Internal Revenue Service (IRS), which provide that (1) the management contract has a term of less than five years (including renewal options); (2) at least 50% of the payments under the contract are on a fixed-fee basis; (3) no payments are based on a share of net profits; and (4) the contract is subject to cancellation by the owner at the end of any three-year period. This is one of the most common forms of public-private partnerships—a private party is brought in to maintain and operate a public facility, while the governmental unit gets to exercise closer control and involvement by owning and financing its own project.

Private Activity Bonds

As the technology of MRFs and disposal facilities becomes more complex, however, local governments may not want the responsibility or expense of owning such a project. Government officials are still anxious, though, to encourage private industry to undertake such projects, so they are usually willing to provide assistance by issuing tax-exempt bonds.

In those cases where a private company will own the project, or will have a more extensive involvement in its operation and maintenance, it will instead be necessary to use the other type of tax-exempt bonds referred to as "private activity bonds." The Internal Revenue Code provides a series of tests that bond attorneys must apply in determining whether more than 10% (or 5% in some cases) of the proceeds of a bond issue will be used in the trade or business of a private party and thus be treated as a private activity bond issue.

Private activity bonds are subject to more restrictions and limitations than governmental use bonds. For example, there is a limit to the amount of private activity bonds that may be issued each year in a state, based on the population of the state. The state ceiling for any calendar year can be one of two figures—an amount equal to $50 multiplied by the state's population, or $150 million—whichever is greater. The issuer of bonds for a private recycling facility must obtain a portion of the state's "volume cap," competing with the sponsors of other large projects that are seeking tax-exempt financing during the same year.

Exemption from the volume cap limitation is possible if a solid waste disposal project is owned by a governmental unit. The private company that leases or operates such a facility under a long-term contract must, however, comply with certain conditions of the Internal Revenue Code to qualify for this exception.

Other Tax Limitations

- Interest on private activity bonds is included in the calculation of the alternative minimum tax for individuals and corporations, so that such bonds must offer somewhat higher interest rates than governmental use bonds, to attract bondholders.
- No more than 25% of the proceeds of private activity bonds may be used to acquire an interest in land.
- Private activity bond proceeds cannot be used to finance the acquisition of used property, unless a certain amount of rehabilitation occurs.
- No more than 2% of a private activity bond issue may be used to pay the costs of issuance of the bonds.
- Property financed with private activity bonds must be depreciated on a slower, straight-line basis.
- Certain excess investment earnings must be paid to the federal government as "arbitrage rebate."
- The governmental unit where the project is located must approve the bond issue after publishing notice and holding a public hearing.
- No private activity bonds may be guaranteed by the federal government.
- Each new bond issue must be reported to the IRS, the bonds must be in registered form, and the average maturity of such an issue is limited to 120% of the average reasonably expected remaining economic life of the facilities.

Obtaining Tax-Exempt Financing

A private company (the Company) seeking to obtain low interest-rate financing for a recycling or solid waste disposal project should start by selecting a law firm that specializes in municipal finance to serve as bond counsel. Bond counsel can explain the tax rules discussed earlier and provide assistance with the procedures listed below.

After the Company provides a detailed description of its recycling or solid waste disposal plan, bond counsel should make an initial determination that the project will qualify as a "solid waste disposal facility" under the Internal Revenue Code. This generally includes those facilities that collect, process, or dispose of waste material that starts out in a solid or semisolid form, plus has no value at the beginning of the process. Although recycling equipment may eventually recover valuable products, such projects usually qualify under the standards set forth in rulings and regulations of the IRS.

Bond counsel should then determine what state agency or local government can act as issuer of the bonds (government issuer), and which state statutes will apply to the financing of the project. If more than one governmental unit is available, the Company should determine which entity to ask to be the governmental issuer. The Company should then contact an official of the governmental issuer to obtain the necessary application form and to convince the governmental issuer to assist the Company with this financing.

Bond counsel should also prepare the form of a preliminary bond resolution to be adopted by the governmental issuer. IRS rules prohibit the Company from using Bond proceeds to pay for any major expenditure that it incurs on a project prior to the adoption of this inducement resolution—which signifies the governmental issuer's intention to issue bonds for the project in the future.

An investment banking firm should then be selected to advise the Company on structuring the transaction and to later market the bonds as the underwriter. After analyzing the financial standing of the Company and the plans for the project, the underwriter will suggest a combination of one or more of the following sources of repayment or security for the bondholders.

- *Guaranty* of the Company, if the Company is financially strong.
- *Letter of credit, surety bond or bond insurance* purchased from an institution with a high credit rating, to guaranty payment of the Bonds as a "credit enhancer."
- *Project financing*, where the bondholders are secured only by the revenues of the project (requiring an analysis of the construction risks, fluctuations of market values for recyclables, expected volume usage and per-ton charges, operation and maintenance expenses, and projected revenues).
- *Mortgage* of the project, to further secure the bondholders and credit enhancers.
- *"Put or pay" contracts* with municipalities, which agree to deliver specified annual amounts of waste to the project and to pay for a set quantity, despite any shortfalls that occur.
- *Flow control ordinances,* whereby municipal users assure a steady stream of waste by requiring all haulers in their jurisdiction to deliver waste to the project.

The Working Group

The underwriter, the Company, and bond counsel should then determine if progress on the project is far enough along to warrant financing transactions by reviewing such factors as the status of construction and environmental permits, title to the site, a feasibility study if one has been conducted, waste supply contracts in hand, and any provisions that have been made for disposal of the residue.

If this working group is convinced of the feasibility of the project, then bond counsel may begin drafting: the Company agreement with the governmental issuer; the indenture, which provides the detailed terms of the bonds; and the bond purchase agreement which provides for the sale of the bonds, the guaranty and the mortgage, if any. The bond counsel should distribute these drafts to all parties for review.

Next, bond counsel should discuss with the Company and the governmental issuer the method and timing of obtaining an allocation of the state ceiling, or volume cap, for private activity bonds. If desired, a meeting to review the instruments and terms should be held and a revised draft of each instrument should be sent out by the preparer.

After preparing a project report, the bond counsel should determine the total cost of the project and which components qualify for financing. Such a report describes the solid waste to be disposed and the purpose, costs, and function of each portion of the project. The Company should describe the status of all governmental permits needed in order to construct and operate the project.

Final Steps

When it is time to offer the bonds to private investors or to the public, the underwriter will use an official statement or other disclosure documents to describe the project, the financing arrangements, the credit-worthiness of the company, and other aspects of the financing.

If the investors will be relying solely on the revenues of the project for repayment, a professional feasibility study that illustrates how expected revenues from material sales and tipping fees will pay debt service, and operation and maintenance expenses will be needed. The underwriter may suggest obtaining a rating of the bond issue from a national rating agency, to provide further guidance to potential bondholders of the degree of safety of their investment and the likelihood of repayment.

After the instruments are in substantially final form, the Company's board of directors should take the necessary action to approve the instruments that will be signed by the Company in the transaction. The governmental issuer's governing body will adopt proceedings to authorize the bonds and, after adoption, will execute the bond purchase agreement. A public hearing must be conducted, after no less than 14 days' published notice. Final documents and closing papers will be executed by all parties, and the closing will take place approximately one or two weeks after the execution of the bond purchase agreement.

Cooperation

The expansion of recycling operations requires a great degree of cooperation among various parties, including: local governments, which use their local siting approval and flow control powers; private industry, which provides the latest recycling technologies and markets for recovered materials; and state and federal regulatory agencies, which establish overall recycling goals and standards to protect the environment.

Tax-exempt bonds issues can significantly reduce the cost of constructing recycling facilities. These financing methods will generally be presented by a team of professionals, including financial advisors, underwriters, and the bond counsel. Such experts will explain the range of options that are available to help communities meet their solid waste disposal and recycling goals in an efficient and cost-effective manner.

****

WTE AND RECYCLING: UNDER ONE ROOF

By *Randy Woods*, managing editor of *Waste Age* and associate editor of *Recycling Times*, Washington, D.C.

Throughout the lengthy and laborious siting process to gain approval for a waste-to-energy (WTE) facility, it is often necessary to debunk misconceptions about the industry.

For example, WTE and recycling have been seen as opposites—one destroys, one reuses. Since mass-burn WTE plants accept municipal solid waste (MSW) with only minimal presorting, environmentalists and some government officials claim that a conflict of interest exists between WTE and recycling. WTE facilities undermine recycling programs, they say, by taking away too many potential recyclables. This line of thinking has contributed to long siting battles, increased NIMBYism, and incinerator moratoria.

But today's WTE facilities are proving that WTE and recycling are not mutually exclusive industries. Nearly every energy-generating incinerator has some kind of front- or back-end metals recovery system, and many are sited in communities that have their own recycling programs.

According to a recent study by the Integrated Waste Services Association (IWSA), many communities that include WTE facilities have met and often exceeded the national average of recycling 17% of the waste stream. The Long Island, New York, communities of Babylon, Islip, and Hempstead—each with WTE facilities—report recycling rates of 30–35%; both Hillsborough County, Florida, and Hennepin County, Minnesota, are recycling more than 40% since the advent of WTE plants, the IWSA study states.

To take this cooperation a step further, some new plants are being designed with more extensive preprocessing lines, and even full-fledged materials recovery facilities (MRFs) built right in. As WTE plans are drawn up in some areas, recycling, composting, and landfilling are beginning to play a role almost as important as the waste combustion itself.

More Than Enough in Detroit

While the dividing line between WTE and recycling is blurring, it is clear that no one combination works best for every community's individual waste management program. Differences in collection methods often dictate what type of WTE system would best fit in with each municipality. Mass-burn facilities, for instance, usually are most cost-efficient in conjunction with a strong recycling program. Noncombustible recyclables—such as glass and metals—and yard wastes (with its high moisture content), which lower the Btu value of the boiler feedstock, can be removed in most recycling programs. Refuse-derived fuel (RDF) facilities, however, have more extensive front-end screening and can remain cost-effective in areas with smaller recycling programs.

Such is the case for the metropolitan area of Detroit, Michigan. According to Mike Brinker, general manager of the Greater Detroit Resource Recovery Authority (GDRRA), there is no curbside collection of residential recyclables in the city. The three-county area that comprises GDRRA relies, instead, on drop-off centers and mobile units in the area for recyclables collection.

"We also have an excellent 10-cent bottle bill in operation since 1978, which keeps most of the glass and aluminum containers out of the waste stream," Brinker says. "Some newspaper and plastic is accepted at recycling centers, but they make up a very minor percentage. Since they are such low-value commodities, it's more cost-effective to send them to the RDF plant."

That plant is the Greater Detroit Resource Recovery Facility, located on a 17.5-acre site within the city limits, four miles from downtown. With a total throughput capacity of 3,300 tpd, the facility is the largest RDF operation—and one of the largest WTE facilities—in the country. After 13 years in the making, from conceptual planning in 1976, to groundbreaking in 1986, to uninterrupted operation in the summer of 1989, the plant accepts all of Detroit's residential refuse (90% of the total daily tonnage), and some commercial waste from private haulers, Brinker says.

Though the facility—constructed and operated by ABB Resource Recovery Systems (Windsor, Connecticut)—is designed to handle 4,000 tpd of MSW on its three processing lines and boilers, permit constraints allow only two lines to operate at once, Brinker says. As a result, no more than 2,200 tons of refuse can be burned in a day. "Adding up the populations of the three [GDRRA] counties—Wayne, Oakland, and Macomb—about one-fourth of the state's population lives in this area, and they produce about 40% of the state's solid waste," he says. "The bottom line is, whatever cannot be recycled from this waste stream goes to our facility. From such a wide area, there will always be more than enough waste generated to economically run this facility that cannot feasibly be recycled."

The Proto-MRF

But the Detroit RDF plant does not leave the job of recycling to others. The three lines of the processing facility, adjacent to the boiler building, contain equipment closely resembling that of an average MRF. "Actually, a lot of the MRFs we see today are sort of a hybrid of the RDF process," Brinker adds.

When the Detroit Department of Public Works refuse trucks enter the facility, their contents are dumped on a tipping floor, where inspectors watch for any obviously hazardous materials that may be included. A Caterpillar 9080C loader then shoves the refuse into an inspection station, where overhead grapples remove hard-to-process items like white goods, propane tanks, mattresses, etc. These materials make up about 1.5% to 2% of the incoming waste, says Larry Evans, general manager of the facility.

After initial inspection, the rest of the bulky waste is sent by Rexnard conveyors to the low-speed CE Raymond primary shredder, or flail mill, Evans says. The shredder's half-inch-thick hammers rotate through the refuse, breaking bags and reducing the waste to pieces from 6 to 12 inches in size. The chunks are then passed under a modified, rotating Grizzly magnetic drum, which removes about 90% of the ferrous fraction. As the recovered scrap is carried away, the metal is further cleaned of combustible material by jets of compressed air in what Evans calls an "air chute." The light fraction that is blown off the metal is then sent back to the main line. "All together, we usually pull about 5% to 5.5% [of the waste stream] off as ferrous scrap," he says.

The remaining material is sent through a two-stage trommel with 1-inch- and 4-inch-diameter holes. "We've found that the less-than-1-inch material is usually low in Btu or is noncombustible, like rocks, broken glass, and dirt." Evans says. "That fraction is taken away as residue and makes up about 12% to 18% of the amount of refuse we process. The less-than-4-inch material is mostly paper of the proper size for combustion." Oversize material that passes through, such as larger pieces of cardboard, are sent through a Williams secondary shredder and reintroduced to the trommel screen, he adds.

Savings, Motown-Style

Once the fuel has been processed to this homogeneous, 1–4-inch size, it is ready to be taken by Rexnard and Wolf conveyors to the combustion chamber, where it is burned at 825 degrees F. "RDF burns differently than mass-burn fuel," Evans explains. "The lighter RDF has a higher Btu content and burns in suspension, rather than on the grate, so it burns more evenly and completely."

The heated boilers produce about 387,000 pounds of steam per hour, which is sold to the Detroit Edison Company for distribution to central heating units in the downtown area, about 3.5 miles away. The steam from the facility accounts for about 60% of the utility's needs. About 1% of the energy produced is also sold to Detroit Edison in the form of electricity. About 15% of the material that enters the boilers is disposed of as ash in a monofill, Brinker says.

Last July, the Detroit plant produced 183 million pounds of steam, 10,172 kilowatt-hours of electricity, and had enough left over to power itself, he says. From the nearly 75,000 tons of waste accepted that month, about 2,150 tons of ferrous scrap were removed and sold to Luria Brothers in Chicago, as well as to other local markets. "Being in Detroit, we seem to always have this association with cars," he says. "So, since we estimate it takes about one ton [of scrap ferrous] to produce the average car, we like to say we've saved enough to produce 2,150 cars for that month. We also saved 68,000 barrels of oil or 16,000 tons of coal from the energy we produced."

While most of the other residue that is removed is sent to a landfill, Evans says there are still possibilities for recycling this fraction. "The city has done some experimentation over the years with pulling out some of the inert by-products for reuse or composting the organics," he says. "But so far the results have not been advantageous for the city."

Brinker agrees that the RDF system leaves a lot of room for expansion into more sophisticated recycling efforts, but at the moment, he has his hands full with the ever-changing emissions regulations. Electrostatic precipitators and lime slurry injection treatments currently being used keep particulate, sulfur dioxide, and dioxin emissions well under current standards for operating facilities. But, in order to achieve long-term compliance with new Clean Air Act standards, the facility is retrofitting its boilers to include baghouses and scrubbers. "[ABB] already has these controls in place in similar RDF plants in Hartford, Connecticut, and Hawaii," Brinker says. "We're extremely confident we will achieve compliance with standards, not just for current facilities, but for new ones."

The retrofit for the first boiler is "significant," Brinker says. "It's a five-year adventure in engineering, getting it all done on 17.5 acres," he says. "Once we have all the construction finished [sometime in 1996], maybe we can go after some other recyclable materials and pull more out on the tip floor. When things calm down later on, we can 'play' with the waste a little more."

Though the GDRRA ultimately chose ABB's RDF system for its superiority over mass-burn in terms of steam generation, Evans sees RDF as an added benefit in terms of recycling, as well. "The RDF system can more easily be expanded into a MRF operation since the method of processing is similar," he says. "With MRFs picking up across the country at this point, there's generally a trend these days toward more complementary systems."

More Than Just Metal

The flexibility of the RDF system is working well in Detroit and other areas, but it does not necessarily mean the end of mass-burn facilities. Rather than having one company build and operate a processing facility and WTE plant in one, some municipalities, such as Bucks County, Pennsylvania, are opting for a recycling/WTE partnership that shares the burden with two companies—one markets what it can and the other mass-burns the residue.

In Pennsylvania, there is no bottle bill to help remove containers from the waste stream, so the state must rely on its 25 operating MRFs; in contrast, the bottle-bill state of Michigan has just five operating MRFs, according to *Waste Age's* 1992 Directory of Materials Recovery Facilities. In May 1992, the Pennsylvania Department of Environmental Regulation (DER) issued final permits for the construction of a MRF/mass-burn WTE combination that will accept MSW from Bucks County and the neighboring city of Philadelphia.

While RRT/Empire Returns Corporation currently runs a 45-tpd MRF owned by Bucks County, the facility only accepts old newspaper (ONP), office ledger, clear glass, and aluminum. In the new proposal, Wheelabrator Environmental Systems, Inc. (Hampton, NH), will own, build, and operate a 1,600-tpd, two-unit, mass-burn plant, which will include a MRF on site, operated by Otter Recycling (Bristol, PA). In addition to the regular commodities handled by MRFs, such as ONP, glass, aluminum, ferrous, high-density polyethylene (HDPE), and polyethylene terephthalate (PET), the Otter facility will include old corrugated cardboard (OCC), mixed paper, and at later stages, high-grade paper.

"We're going full steam ahead," says Richard Felago, project manager and Wheelabrator vice president. "We began construction on site in early August and the stack is on its way up."

The $200-million facility, located in Falls Township, is scheduled to come fully on-line in the summer of 1994, but just burning garbage is not Wheelabrator's only priority. The MRF should be operating about a year before any waste is burned, Felago says.

"This is truly a unique facility in that it is the first time ever that a merchant [WTE] plant and MRF have been planned on the same site," he says. "We're really going out on a limb on this one." By establishing new markets well in advance of WTE startup, he says, Wheelabrator plans to easily meet Pennsylvania Act 101, which mandates 25% waste reduction and recycling by 1997. Drop-off capacity for recyclables, including button batteries, will also be provided through lower Bucks County and northern Philadelphia, he adds.

Loading the Bases

Then, there are those who ask, if you can site a MRF and a WTE plant together, why not do it all? ABB Resource Recovery Systems, learning from its Detroit facility, is trying to answer that question with the West Suburban Recycling and Energy Center (WSREC) project in Summit, Illinois, just west of Chicago. After negotiating with the Village Boards of McCook and Summit for 18 months, ABB got the green light for construction of its potential grand slam—a combination transfer station, MRF, MSW composting facility, and RDF plant.

Fitting snugly into an urban, 36-acre site, the $200-million project is scheduled to accept about 2,050 tpd of MSW from the Chicago metropolitan area, pending siting and permitting approval. Once the site has been decided, in either Summit or McCook, ABB and co-developer, Kirby-Kaufman, Inc. (Springfield, IL), will apply for permits from the Illinois EPA for a planned startup in July 1996. Over the next three years, however, WSREC will begin processing waste in four distinct phases.

In the first phase, a transfer station will be set up to separate recyclables from incoming MSW. The temporary facility, expected to open in January, will arrange for the shipment of recyclables to local markets and nonprocessible waste to area landfills. By July 1993, ABB officials hope to have the second-phase MRF on-line to process and market mixed loads of OCC, ONP, magazines, phone books, glass, aluminum, ferrous metals, HDPE, PET, and polyvinyl chloride (PVC). The 250-tpd Kirby-Kaufman MRF is being planned with possible blue-bag or curbside-separation collection programs in mind, officials say, and will also process white goods for recycling.

The unmarketable waste will then be sent through phase three—the first MSW composting facility in the state of Illinois. ABB expects about 400 tpd of mixed organic wastes and source-separated food and yard wastes to enter the facility when Kirby-Kaufman completes construction in 1995. With the combination of the closed, in-vessel compost system and the fully operational MRF, ABB and Kirby-Kaufman predict a 35% to 40% waste recycling rate.

In the final phase, the remaining 1,400 tpd will be processed once more at ABB's RDF plant. Similar to ABB's Detroit facility—and two others in Hartford, Connecticut, and Honolulu, Hawaii—the WSREC RDF system of trommels, shredders, and magnets will pull out any residual metals, which will be sent back to the MRF for sale to end-users. Between 1,100 and 1,200 tpd of clean RDF will be produced and burned, according to ABB. The facility's steam turbine generator will produce 42–45 megawatts of electricity; 20% will be used to power the facility while the rest will juice 20,000 homes in the Commonwealth Edison power grid.

This type of fully integrated processing is the ultimate step in what some believe is an inevitable symbiotic relationship between WTE and recycling. "There's a growing move towards more front-end processing in the waste-to-energy industry," says ABB's Evans. "I think RDF plants will be an evolutionary change we'll see through this decade."

"When you site [WTE] facilities, there's always a lot of disproving that needs to be done," adds GDRRA's Brinker. "Opponents will always be saying 'the sky is falling!' It's up to us to remind them that the sky was there yesterday, it's up there today, and it will stay there tomorrow. Some also say that waste-to-energy is against recycling—that's just another myth."

****

PACKAGING—REDUCE, RECYCLE, OR WHAT?

By *Chaz Miller*, recycling manager for the Environmental Industry Associations, Washington, D.C.

Terry Bedell should be ecstatic. As manager, environmental products, for the Clorox Corporation (Oakland, CA), he has a good story to tell about Clorox's success in source reduction and recycling. Since 1988, Clorox has successfully reduced its packaging by 30 million annualized pounds. In addition, Clorox uses 22 million pounds of recycled material in its packages. On an annual basis, Clorox has reduced its use of virgin packaging materials by 52 million pounds, or more than 15%.

Source reduction successes include eliminating 18 million pounds of glass in Clorox's Hidden Valley Ranch salad dressing bottles and 7 million pounds of linerboard in redesigned corrugated containers used to ship Clorox products. Clorox also realized smaller savings in plastic and other glass containers. "The big source reduction surprise was in glass bottles, and it was simply a matter of smoothing out radii and other minor redesigns," Bedell says.

Recycled content includes 8 million pounds of plastic (which was expected to double by the end of 1994). In addition, Clorox's glass bottles average 25% recycled content and its corrugated containers average 35% recycled content.

So why isn't Bedell ecstatic? Because, he says, while legislators congratulate Clorox for its successes, they still want to pass mandatory recycling and source reduction requirements that will "lock us into a legal structure that doesn't make sense."

Bedell's dilemma is a common one confronting packagers and retailers these days. Packaging is often singled out as a solid waste villain. Much of the Resource Conservation and Recovery Act reauthorization debate in 1992 focused on various ways to reduce the amount of packaging destined for disposal.

Packaging is good, bad, or neither, depending on who you are. To a packager, good packaging ensures food safety and the survival of the product in the distribution system. To a hauler, packaging is just one of many materials placed on the curbside for disposal. To a recycler, packaging is a primary raw material, but one that doesn't always pay its bills.

But what is packaging and why do we use it? Should it all be recyclable, or does it make sense to simply promote "less" packaging, even if the result is unrecyclable packages?

What Does Packaging Do?

Packaging comes in many shapes, forms, and materials. Bob Testin and Pete Vergano, packaging professors at Clemson University (Clemson, SC), list four functions of a package—containing the product, protecting it, informing the consumer of what is in the package, and making the product easy to use.

Consider toothpaste. It is usually sold in a tube that comes in a box. The tube contains and protects the toothpaste. It also informs the consumer on how to use the toothpaste and what is in it. The tube also provides the convenience of squeezability and avoids the inconvenience of mixing a powder with water. The tube is "primary" packaging.

"Secondary" packaging, the box, protects the tube and helps prevent theft. It also has a marketing and information function. These boxes are often made from recycled paperboard. As a result, they provide both a packaging function and a market for wastepaper.

Finally, the tube in the box is shipped along with many other tubes in boxes in a corrugated box which protects the product. That container is probably wrapped together with other corrugated boxes by shrink-wrap and carried on a wooden pallet to make transportation more efficient and less costly. Transportation packaging is often called "tertiary" packaging.

How Much Is Out There?

Consider the following:

- We are drowning in packaging.
- Packaging is the largest and fastest growing part of the waste stream.
- All packagers design packaging to be as small, lightweight, and efficient as possible.

Now, are these statements true or false? Of course, that's a trick question. Depending on how you want to look at it, the amount of packaging has skyrocketed or it has stayed flat as a percentage of municipal solid waste (MSW).

According to the U.S. EPA, the amount, by weight, of packaging waste (before recycling) increased by 136% from 1960 to 1990 (27.3 million tons to 64.4 million tons). However, the amount sent to disposal only increased by 96% (24.2 million tons to 47.4 million tons). At the same time, total MSW generation grew by 123% and disposal by 98%.

As a percentage of MSW, packaging only increased slightly, from 31.1% to 32.9%. More importantly, packaging actually declined slightly as a percentage of MSW sent to disposal. In 1960, packaging was 29.5% of solid waste discards. In 1990, it was 29.2% (see Table 7.1). Perhaps it's fair to say that the packaging grew no differently than solid waste, as a whole.

Interestingly, packaging outgrew the population (it increased 39%), but it didn't outgrow the American economy (as measured by the gross domestic product in constant dollars, it grew by 148%).

Why is packaging's share of MSW remaining flat? After all, aren't we drowning in a rising tide of packaging? According to Marge Franklin, president of Franklin Associates (Prairie Village, KS), a number of reasons come into play. More packaging is recycled and heavier packages (glass and steel) are being replaced by lighter packages (aluminum and plastic).

Franklin also points out that solid waste has changed over the years. Both durables (appliances, furniture, tires, etc.) and nonpackag-

Table 7.1 Weight and Volume of Landfilled Packaging, 1990

	1990 Discards (million tons)	Weight (% of total)	Landfill Volume (mil cu yd)	Volume (% of total)
Glass	9.3	5.7%	6.6	1.6%
Steel	2.3	1.4%	8.0	1.9%
Aluminum	0.9	0.7%	5.9	1.4%
Paper & paperboard	20.6	14.0%	53.9	13.1%
Plastics	6.7	4.1%	41.2	10.0%
Wood	7.5	4.6%	18.7	4.5%
Other	0.2	0.1%	0.4	0.1%
Total packaging	47.4	29.2%	134.8	32.7%
Nonpackaging				
Durables	24.8	15.3%	95.3	23.1%
Nondurables	43.1	26.6%	125.3	30.4%
Food wastes	13.2	8.1%	13.2	3.2%
Yard trimmings	30.8	20.0%	41.3	10.0%
Misc inorganics	2.9	1.8%	2.3	0.6%
TOTAL MSW	162.2	100.0%	412.2[a]	100.0%

[a]Volume total derived by adding the individual factors. Actual landfill density may be considerably higher.

SOURCE: Characterization of Municipal Solid Waste in the United States, 1992 Update, U.S. Environmental Protection Agency, Office of Solid Waste, 1992.

ing paper and plastic products have increased greatly in the last 30 years. Remember that record player you replaced with a cassette player that you replaced with a compact disc player? What about all those vinyl records you no longer need? Those cassettes? And what about the explosion in the use of paper in spite of all the hype about the paperless office? Since 1960, durables in the waste stream have increased by 175% and nondurables by 184%. Packaging grew, but not by that much.

Surprisingly, perhaps, transportation packaging is the most common form of packaging. More than half, or 33.3 million tons of the 64.4 million tons of packaging generated in MSW in 1990, is transportation packaging. By contrast, primary packaging is slightly less than one-third of generated MSW, with the remainder being secondary packaging.

Table 7.2 1992 Packaging Recycling Rates

Aluminum cans	67.9%	1,100,000	tons
Glass bottles	33.0%	2,400,000	tons
Steel cans	40.9%	970.000	tons
Plastic: bottles	18.8%	407,000	tons
Plastic: all packaging	6.5%	470,000	tons
Paper: corrugated	59.0%	15,300,000	tons[a]
Wood pallets	5.0%	400,000	tons[b]

[a]Includes pre-consumer scrap.

[b]1990 data.

SOURCES: Aluminum Association, Can Manufacturers Institute, Institute of Scrap Recycling Industries, Glass Packaging Institute, Steel Recycling Institute, American Plastics Council, American Forest and Paper Association, U.S. Environmental Protection Agency.

Two-thirds of transportation packaging is corrugated boxes. Most of the remainder, by weight, is wood crates and pallets, which generated 7.9 million tons of MSW in 1990. "America uses disposable wooden pallets to a greater extent than any other country," according to Jim McCarthy, senior analyst, Congressional Research Service (Washington, DC). Only corrugated boxes and glass bottles generate more packaging than wood packaging.

By material, paper is easily America's package of choice. More than 50% of all packaging is paper. Glass is the next highest at 18.5%. Wood, plastic, and metal packaging follow (see Table 7.2).

Collection and Recycling

Most packaging, like most solid waste, is still disposed of through incineration or landfilling. According to the EPA, 33.4 million tons, or 26% of all packaging, was recycled in 1990. Half of the 33.4 million tons of MSW that was recycled or composted in 1990 was packaging.

Packaging's recovery rate was higher than any other MSW category tracked by the EPA. Aluminum cans and corrugated boxes have the highest recycling rates, while corrugated boxes and glass bottles lead in tonnage. The only nonpackaging product with high recycling numbers is newspaper.

While some materials claim particularly high rates for individual types of packages, those numbers are of dubious value and accuracy.

After all, many programs collect polyethylene soft drink and custom bottles. How do the counters distinguish between the two types? The same questions could be asked for recycling rates for glass beverage and food bottles.

Collection of packaging varies. Primary packaging is commonly collected by curbside recycling programs. Drop-off and buyback centers and deposit systems also are used for primary packaging. While some residential recycling programs collect secondary packaging, the majority do not. Most of the tertiary packaging collected for recycling is corrugated boxes from businesses, not residences.

According to a Waste Recyclers Council (WRC, Washington, D.C.) study [see Chapter 3], commingled containers cost an average of $83.36 a ton to recycle (with a range of $40.76 to $146.29 per ton at the 10 facilities studied). The average cost of processing a ton of newsprint was only $33.55 (with a range of $20.43 to $55.93). Processing cost for packaging was increased by its low volume and high labor cost.

A WRC study of collection costs (see Chapter 3) showed glass and newspaper to be the least expensive materials to collect, largely because of their weight. Unlike the processing study, which looked at actual operating costs, the per-material costs in the collection study were based on a computer model of collection conditions in a "typical" suburb.

With an average revenue for recyclables below $40 per ton, the cost of collecting and processing is clearly greater than the revenues from the sale of recyclables. However, recent studies in Washington and Minnesota have shown that recycling programs in states with high disposal costs can be cost-effective.

Source Reduction

Because source reduction is at the top of the solid waste management "hierarchy," many think it should have priority. However, Bette Fishbein, senior fellow with INFORM, Inc. (New York City) a nonprofit research organization, points out that while most states put source reduction at the top of their hierarchy, when it comes to budgets and full-time positions, they "skip over source reduction."

Why does source reduction get short shrift? One reason is simple. As Bedell puts it, "recycling is easy to measure, source reduction is not." Packagers are frustrated because they find it virtually impossible to get credit for source reduction successes.

Another reason is that source reduction is hard to define. INFORM defines source reduction as "reducing the amount and/or toxicity of waste actually generated." Most of the other published definitions are similar.

But what does this mean? Is it using less materials? Is it making packages smaller? Is it designing a package that is lighter, but takes up the same or more space in a collection truck or a landfill than competing, heavier packages? Source reduction also carries its own controversies. Plastic, after all, is lighter than competing forms of packaging. Shouldn't packagers get credit for source reduction simply by using plastic instead of heavier, competing packages? Pam Murphy, author of *The Garbage Primer*, argues that effective source reduction "should include weight and volume reduction" because one without the other may not be sufficient.

According to EPA data, in 1990 plastic packaging was 4.1% of the weight of landfilled MSW and 10% of the volume. Aluminum and steel also contribute more to landfills by volume than by weight, while paper and glass contribute more weight and less volume (see Table 7.2).

In the marketplace, source reduction has a mixed record. Jacquelyn Ottman, president of Ottman Consulting (New York City), believes consumers accept source-reduced packaging when they see a benefit (as in concentrated laundry detergent) or when it slips by them. Examples of the latter include the steady lightweighting of aluminum cans and eliminating pieces of a package. Ottman cites a package for dryer sheets that eliminated the dispenser roll. Sheets now simply pop out one at a time—a source reduction benefit that simply slipped by consumers.

A number of products illustrate the ups and downs of source reduction. Concentrated products that allow less product to be used than in nonconcentrated products, have been wildly successful in laundry detergents. Concentrates now dominate the laundry detergent market, according to Tom Rattray, associate director of Environmental Quality for Procter and Gamble Company (Cincinnati).

Why are they successful? Rattray points out that they do not require refills or any other action by the consumer. They do the same job with less material. Ottman notes that retailers like the concentrates because the packages use less shelf space and allow more product to be sold in the same amount of space formerly occupied by the nonconcentrates.

Frozen, concentrated orange juice, however, has not had a happy decade. It has steadily lost market share to orange juice in ready-to-serve packages (cartons, bottles, etc.). In 1986, ready-to-serve orange juice claimed over half of orange juice sales and the gap has widened ever since. This is due to ready-to-serve's convenience (it doesn't require mixing), along with improved economics and product value, according to Harry Teasley, former president of Coca-Cola Foods (Atlanta).

In many ways source reduction and recycling may not be totally compatible. Aluminum cans and recycled content in corrugated boxes provide two examples. Continued improvements in lightweight cans (cans have gone from 21 cans per pound to 29 cans over the last 20 years) and low aluminum prices may discourage recyclers from collecting cans. As for corrugated boxes, higher levels of recycled content can make the box heavier. As fibers get shorter, more material is needed to ensure that the box meets compression, stacking strength, and burst tests.

Perhaps a more important question is: If a package has source reduction advantages but is hard to recycle, which is more important, recycling or source reduction? Intuitively, it would seem that source reduction is more important. However, is source reduction enough? How do you prove a material has fewer environmental impacts? A source-reduced package may well carry other environmental costs. Life-cycle analysis is the attempt to assess these issues and decide which packages are best for which applications. However, virtually all life-cycle analyses invariably get attacked for overemphasizing one attribution and underemphasizing a drawback. With no agreement on how to conduct a life-cycle analysis, source reduction will continue to be honored more in the breach than the observance.

****

SCAVENGERS: A BEHIND-THE-SCENES RECYCLING BATTLE

By *Jennifer A. Goff*, managing editor of *Recycling Times*, Washington, D.C.

Though it may not garner a spotlight as the number-one issue facing the recycling industry, the theft of recyclables is a subtle and elusive vexation that is costing haulers and recyclers thousands of dollars each year.

Scavenging is not a new dilemma for the industry. "The problem has plagued our program since it started," admits Brooke Nash, executive director of Solana Recyclers (Encinitas, CA). But aside from an occasional seminar on the topic, not much attention has been devoted to the problem, despite the fact that poaching deprives curbside programs of revenues that help offset operating costs.

"When we lose revenue, we lose some of our budget, so it is detrimental. We're losing about $10,000 per month in Berkeley," says Barbara Hall, recycling manager for the Ecology Center (Berkeley, CA).

"We estimate that we lose about 20% of our aluminum at the curb, and that costs us around $20,000 to $30,000 dollars per year depending on the value of aluminum," Nash echoes.

In some locations, losses due to scavenging have sabotaged the efficacy of curbside recycling as a whole. The city of San Diego, for example, has been hit particularly hard. For the first half of 1994, approximately 2,895 bins holding recyclables were stolen, costing the city about $17,370 to replace, according to statistics provided by San Diego's Environmental Services Department. "People become annoyed when they see [scavengers] going down in station wagons picking up the materials they've sorted," says Maureen Dixon, recycling specialist for the City of San Diego. "Increased scavenging causes decreased recycling efforts. We're losing $15,000 in equipment [i.e., containers] alone—and that's not even counting the contents," she adds.

Curbing the Problem

Addressing the issue of curbside theft is difficult for several reasons. One of the most troublesome is the fact that incidences of scavenging are often sporadic, making it hard for haulers and police to identify and catch poachers. Several states and cities have placed anti-scavenging ordinances on the books, but the problem still persists. Residents who see poachers often will report them to the police, but by the time a law enforcement officer arrives on the scene, the perpetrators are usually gone. "Unfortunately, calls like that get pushed to the bottom of the barrel," Nash says.

"The actual scavenging has to be witnessed," Dixon adds. "So instead of citations, [code enforcement officers] issue a lot of warnings."

In fact, at least one company, Serio-Us Industries, Inc. (Hanover, MD), sells locks for recycling containers to prevent poaching.

However, the illegal aspects of scavenging are not limited to the loss of revenues. In some cases, scavenging incidences have threatened the personal safety of the drivers. "Our drivers can't act as police," Hall says. "There have been a couple of situations where [the scavengers] have gotten aggressive, and it makes it unsafe for our drivers."

In order to both address the safety concerns and try to bridle the extent of the losses, both the city of San Diego and Solana Recyclers promote antiscavenging tactics in their educational materials that are sent to residents. "We [tell residents] what they can do to try and cut back on scavenging—like wait until the morning to put their materials out, and to report license numbers if they see a pickup truck [that is being used to haul materials]," Nash explains. The city of San Diego also suggests placing the bin of newspaper (which has lesser market value) on top of the bin that holds the more sought-after aluminum, plastic, and glass.

"Some haulers take an even more aggressive approach because they say 'Look, this is money going out of our pocket,'" Nash says. "But I did hear about a scavenger pulling a gun on one of the route supervisors. That can get to be a serious problem."

Stealing or Surviving?

The social ramifications of scavenging present other considerations that can complicate the issue. It is not uncommon to see homeless people burrowing through waste containers in order to collect materials that retain some monetary value. For the most part, the public has been reluctant to report these instances to the authorities.

"In some ways, you get into a touchy social issue if [the scavenger] happens to be a homeless person on foot. Residents don't want to deprive them of a few cans that will pay to put food in their mouth," Nash says.

"The homeless scavenging is a relatively small percentage. They've gotten a bad rap on that," Dixon says, adding that the priority for San Diego is apprehending the organized scavengers, even though, by law, all perpetrators must be treated the same.

In some cases, the homeless actually provide a service to recycling, according to Carl Hultberg, president of the Village Green Recycling Team (New York City), a community-based recycling center that organizes homeless individuals to help collect recyclables. The homeless "are doing us a favor—an essential duty," Hultberg says.

"They're not the type that will slash bags open and leave a mess. We have a huge mass of recyclables that need sorting, and...we have a large mass of disadvantaged people who are looking for something to do," he adds.

According to Hultberg, the homeless collect cans primarily off the street, though he admits that they may extract cans from curbside containers. "We would encourage a large [apartment] building to separate its deposit containers and have the homeless pick them up," says Marcia Bystryn, assistant commissioner for the Bureau of Waste Reduction, Reuse, and Recycling for the city of New York. "We don't do anything in a regulatory sense about the [homeless] scavenging. For us, it's primarily a litter issue. We don't want the garbage left all over."

From all accounts, the primary nemesis of haulers is the type of scavenger that pilfers curbsides, not to get a few cents for food, but to make a living. Unlike the homeless issue, the public reserves no sympathy for professional poachers. "It's the people in the station wagons and the trucks that we're really concerned about," Dixon says.

"We were experiencing [illegal] businesses going down the alley emptying our recycling containers into the back of their truck," explains Wanda Wildman, solid waste management inspections administrator for the city of Phoenix.

"If residents see someone driving down the street in a nice pickup truck taking the stuff, they get upset," Solana's Nash says. "They think, 'I'm the one sorting this stuff and the revenue is paying for the recycling of it.' What's really frustrating is that they're taking the containers as well."

Hot Commodities

Other factors exacerbate the scavenging problem, further frustrating haulers and recyclers. While the news of improved markets for recyclables is always good, it also means that the likelihood of the commodities falling victim to scavenging is up as well. "It's safe to say that people will go with what has the most value," says Tom Estes, spokesman for the California Integrated Waste Management Board (Sacramento).

"When the prices go up, we see more theft. We were seeing some theft of glass when the price was up," Nash adds. Besides market value, certain kinds of legislation also may affect the degree of scavenging. In California, for instance, a 1986 beverage container recycling law and a recent rise in curbside collection programs reportedly have made scavenging a more lucrative practice in the state. Though several other states have implemented "bottle bill" legislation, California is the only state in which the containers can be redeemed not only for the deposit, but for the scrap value as well, Nash says.

In July, the value of aluminum was about 80 cents per pound, while the "California Refund Value" was approximately 67 cents per pound. "The scrap value is less than the redemption [value] but [the beverage container law] probably affects [the amount of scavenging] for the general public," Dixon says.

To date, no studies have been conducted that make a direct connection between the beverage container law and the degree of scavenging in the state. "If [California] is hard hit, we have no quantification of that," says Pamela F. Morris, public affairs officer for California's Department of Conservation.

Though a total of six states in the Northeast have implemented bottle bills, scavenging has not been as much of a problem as it has been in California. "I'm sure it happens, but it's just not that widespread," says Michael Alexander, policy analyst for the Northeast Recycling Council (Brattleboro, VT). "I think it has more to do with the local economy [than the beverage container redemption legislation]. If you took away the scrap value and still had the deposit value, you'd probably still have a problem [in California]," he adds.

If You Can't Beat 'em, Co-Op 'em

In South America, scavenging is viewed quite differently than in the U.S. "Scavenging is the basis of recycling in the third world," explains Christopher Wells, executive director of Cempre, a not-for-profit, private-sector association that promotes integrated waste management in Brazil. "We are putting out an educational kit that would allow a volunteer to give a course to a group of scavengers so that they can form a [recycling] co-op."

The course will emphasize health and safety issues and also will help scavengers—who are collecting the materials anyway—to cultivate their activities into a type of business so as to obtain better prices for the materials. "People have lost their jobs, and the idea of the co-op . . . is to get these people to get some self-esteem, self-awareness. The scavenger can earn two to three times above the minimum wage, which is about $80 per month [in Brazil]," Wells says.

Cempre contends that the municipally run programs "work far better when integrated with street scavengers and informal scrap dealers because it reduces their costs," according to a promotional brochure published by the organization. A cooperative in São Paulo, the world's second largest city, supports approximately 250 people who are making a living off of the recyclables they collect from the streets.

"It is something the U.S. could start thinking about," Wells adds.

8 EFFECT OF OVERSEAS REGULATIONS ON U.S. RECYCLING

This chapter at first violates this book's premise that it is not historical in nature but pertinent for current practice. And yet the German packaging recycling mandate affected world markets, influenced world thinking, and created an effect on recycling markets that is still being felt today. And so this chapter includes Waste Age's *first description of the mandate from January 1992 and then follow-up articles on similar approaches tried in other countries, on the effect of the mandate on markets, and of three contradictory verdicts on the mandate. The jury, as the saying goes, is still out.*

GERMANY'S GREEN DOT SYSTEM FOR PACKAGING MATERIAL

By *Adrienne Redd*, assistant recycling coordinator, Bethlehem, Pennsylvania. She has written extensively on waste issues and visited Germany to research this article.

Legislation initiated by Klaus Topfer, the German minister of the environment, and passed by the German Parliament in April 1991 could revolutionize the recycling of packaging. Arguably the most comprehensive law of its kind, the "Verpackungsverordnung," or Packaging Regulation, sets target percentages and dates by which industry must collect and recycle three levels of packaging.

The time table is very straightforward. By December 1, 1991, level one packaging—transport packaging, mostly corrugated cardboard but also barrels, canisters, cases, sacks, pallets, polystyrene, and shrink-wrap—must be accepted back and recycled by manufacturers.

The second level includes distribution or intermediate packaging, including cups, bags, blister pack, cans, barrels, bottles, cartons, canisters, trays, and plastic bags; they must be recycled and collected starting no later than April 1992. The executive summary of Verpackungsverordnung defines this and other intermediate packaging as the materials used to transport and handle goods before they are actually used by consumers. Shoppers can simply take their purchases out of the packaging and either leave the material in the store or can return the material later.

Level three packaging are those that are actually handled by consumers, and includes bottles, tins for coffee (sealed with plastic), cans, tubes (for mayonnaise, mustard, caviar), aseptic boxes, blister pack, and blister bags for wine, milk, and other beverages. By 1995, manufacturers of packaging must collect and recycle 80% of these packages.

To further encourage recycling, the law bans incineration as a means to dispose of packaging.

The reasons for this new legislation may sound very familiar to Americans. Ten years ago, due to pressure from the Green movement, new landfill development in German was severely restricted. Waste-to-energy facilities were proposed as a substitute for landfill disposal, but none are being permitted, again, due to lobbying from the Green movement. Mandated recycling consequently appeared to be the only feasible alternative.

The very strict German standards are expensive. The cost of start-up has been estimated at more than $5 billion, with annual management expenses of $1.2 billion.

At first, packaging manufacturers will carry the cost of the program until they can pass them on to consumers. The structure of the regulation's funding, which will internalize the cost of processing to each piece of packaging, will increase the price of each item. The program will be made publicly visible with a green dot required on licensed packaging. A fee will be required to dispose of packages without a green dot. A goal of the program is to make consumers more aware of excessive packaging and the increased price, and ultimately encourage them to avoid products with excess packaging.

There is a potential weakness to the program—the uncertainty of how recycling is to be implemented. In five articles published in *Mullmagazin*, or "Garbage Magazine," between August 1990 and September 1991, neither proponents nor critics of the Verpackungsverordnung were able to give even a sketchy picture of how up to five million metric tonnes (4.5 million tons) of packaging per year will be collected and recycled, except to say that existing infrastructure will be utilized to as great a degree as possible, and that up to 200 plants will be needed to sort the secondary resources.

Although the implementation of the Verpackungsverordnung is known as the dual system, an executive summary of the regulation issued by the German Trade and Industry Organization (EHK), the equivalent of a Chamber of Commerce, recommends and anticipates that the recycling of packaging will use existing collection systems. This will certainly be true of transport level of packaging, particularly corrugated cardboard. In fact, according to Dr. Ing. Volrad Wollny, project leader at the Oko-Institut, in Darmstadt, 75% of cardboard is currently recycled in Germany. It will probably continue to be recycled via the current collection, brokering, and processing methods.

It is perhaps this great success with recycling, and the fact that much of Germany currently uses a drop-off system, that led the German government to allow just eight months for return and recycling of level one packaging, and another four months for level two packaging.

Industry Response

German legislation, such as the "Abfallgesetz," or trash law, passed in 1972 and later updated, gives government the right to set targets for industry. If industry does not meet the targets, government can then redefine and tighten the goals with regulations such as the Verpackungsverordnung.

When it became clear to industry leaders in 1990 that the Verpackungsverordnung was to be passed, they said they would comply, but wanted to take the initiative in deciding how packaging recycling would be done. At press time, over 400 German and international companies, such as AluSwisse, BP Chemicals, Coca Cola, Nestle, Procter & Gamble, Schweppes, and Unilever, have provided the funding to create the Duales System Deutschland (DSD), a licensing cartel.

The DSD will license every company selling packaged products in Germany, and issue permission to companies to use a green dot, indicating they have been licensed. One to 20 pfennig will be added to the price of individual products; a 0.6-to 12-cent fee. A 50 pfennig mandatory deposit, equivalent to 30 cents, will be placed on all nondairy beverages, cleaning materials, and paint. The fees from the packaging and the deposits will fund a theoretically separate collection system for packaging, thus the name dual system, or dual green dot system.

One retail chain, DM-Drogerie Markets, which has 284 stores in Germany, is voluntarily recycling packaging. According to Herbert Arthen, from the company's public relations department in Karlsruhe, the company made the decision to try recycling boxes in 1989, and set them up in 1990. Currently, 32 stores are participating with a display to which customers can return batteries, paperboard and paper, plastic film, and bubble pack, and mixed paper and plastic. Arthen says that the response has steadily increased,and that a vocational-technical class will soon release results of exactly what weights of packaging have been collected and recycled.

The drugstores of the DM chain are currently the only one with a publicized program, although most groceries currently accept corrugated cardboard to be recycled. Businesses realize that the packaging regulation will soon be implemented and seem ready to respond. For example, Eismann, a European frozen food delivery company, takes back packaging, such as shrink-wrap, pizza and other food boxes, and rinsed ice cream cartons, according to Kurt Goos, an Eismann manager. Other companies are taking related initiatives. Tupperware is promoting the idea of people taking their own containers to be filled at the butcher shop or delicatessen. A less obvious response came from Coca Cola, which had been preparing to introduce disposable PET bottles, and simply didn't because of the Verpackungsverordnung.

Consumer Reactions

In two cities which have the Drogerie Market chains, consumers seem much more aware of the option to return packaging because they can do it now. One homemaker commented, "We have seen advertisements about making less trash on the television. Here is a way we can actually do something."

A father shopping with his small daughter said of returning packaging, "It only makes sense. We must take back bottles and batteries, and the repair shops must take back car oil. Let the makers of the packages take them back and use them over."

The five new states, as former East Germany is now called, seem to have a lower awareness of returning packaging, in part because former East German packaging is less excessive, and because the five new states have grave economic concerns. A young couple from Berlin commented that many stores take back cardboard and other packaging, but that few people currently take advantage of this policy. They also expressed concern that urban German stores have no storage space and no parking lots in which to store packaging before it is collected to be recycled.

No one interviewed, however, said that they would not return packaging because it would be inconvenient. Recycling functions very effectively now with a combination of drop-off bins, bottle deposits and buyback from scrap dealers; there are very few curbside collection programs. One gets a sense that German citizens have a strong "Umwelt Bewusst," or environmental awareness, and would simply see returning packaging as their duty, not to be questioned.

Cost and Financing of Recycling Packaging

In an article in *Mullmagazin* in November 1990 entitled, "Sieger nach Punkten," or "Winner by Points," Bundestag, or Federal Assembly member Franz Heinrich Krey gives a detailed estimate of what the collection and recycling of packaging will cost and how it will be financed. In 1991 and 1992, he estimates that between 1.3 and 1.5 million metric tonnes (1.2 to 1.4 tons) of packaging will be recycled and that the cost will be between 350 and 450 million deutsch marks, or between $210 and $270 million. In 1993 and 1994 he estimates that between five and six million metric tonnes (4.5 to 5.5 million tons) will be recycled, costing the equivalent of $840 million to $1.08 billion. Wollny criticizes Krey's estimates in a later article, asking if these tonnages are materials collected, sorted, or actually recycled and marketed.

According to Krey, however, his estimates do include the total cost, including public information, handling, and plants for sorting materials, and the costs of recycling will be fully covered by the mandatory bottle deposits and licensing fees on packaging. Revenue from licensing assumes an average fee of two pfennigs, or 1.2 cents per piece of packaging.

What is lacking in Krey's article is any description of the physical means by which materials will be collected and processed. Also notably lacking is any comment on supporting markets for the recycled materials, of which there will initially be a glut. It should be noted that Krey is the secretary to the Committee of Packaging and the

Environment, and has a strong tie to industry, for which the Verpackungsverordnung was relaxed.

Implementation

The collection and recycling of packaging, or "das duale Abfallwirtshaftsystem," as it is known to industry, will be funded by the licensing, and will involve a coming together of existing municipal and private collection. Later it will be supplemented by recycling of plastics and other materials which are currently landfilled or incinerated.

Now in place and functioning well are municipally run drop-off bins for glass and batteries; less common are bins for newsprint and aluminum cans. Charities run bins for second-hand textiles. For people willing to take the trouble, steel cans can go to scrap dealers. There is also some municipal pick-up of mixed paper at curbside and apartment complexes.

There is almost no commingled collection of cans, glass, and plastic in Europe. Almost all materials currently collected and recycled are source-separated, so the infrastructure for sorting grades of plastic and paperboard will have to be built. Dr. Wollny, the EHK summary, and other sources concur that at least 200 plants nationwide will have to be built to manually separate secondary materials. There are 60 separation facilities operating now which do not specifically deal with packaging, though they handle industrial discards, such as packing and construction materials. Because recycling has a longer history, the paper markets seem more stable than those in the U.S., though markets will have to be stimulated to offset the glut the mandatory recycling will surely create.

The law requires a second system in the sense that recycling quotas cannot include materials already being recycled. The term 'dual system' is something of a misnomer, however, because the idea behind the DSD is to piggyback on collection and processing. It's not clear whether new plants will be built by private entrepreneurs, by the companies mandated to recycle, or by the government. But once sorting facilities and plastic and paperboard recycling plants are built, other appropriate materials will find their way there to be reclaimed.

Uncertainties in mind, Wollny has nevertheless made some estimates about the German waste stream and the reduction via recycling that the regulation may achieve. According to Wollny, the total waste stream in West Germany, population 60 million, was 35 million metric tonnes (31.8 million tons) in 1985; the household waste was 14 million tonnes (12.7 million tons). In 1985, packaging made up 8 to 9 million tonnes (7.3 to 8.2 million tons) of the household waste stream; this figure does not include transport packaging. Wollny estimates that the Verpackungsverordnung may cause as little as 2.3 million metric tonnes (2.1 million tons) to be recycled for this nation of 77 million in the first year that the regulation is implemented. However, this is on top of an estimated 30% of glass, paper, cardboard, and bi-metal currently recycled. Also, 72% of beer, mineral water, and soft drink deposit bottles are returned now.

Potential Problems

Wollny raises the point that the first objective of solid waste management should be to avoid and reduce waste at the source. The second should be to reuse waste materials, and that recycling should only be implemented after these first two have been thoroughly explored. He asserts that the Verpackungverordnung does little to encourage source reduction and reuse of packaging. However, the EHK summary does note that it is impractical to send back shrink-wrap from Asian textiles, for example, and suggests that the fee on packaging might be encouraged to streamline packaging. The distributor might package 50 shirts to the same destination together in shrink-wrap, rather than wrapping each shirt individually.

The Verpackungsverordnung divides the packaging waste stream into glass, metals, paper and cardboard, plastic, and unsortable residue. Though infrastructure exists to sort the first three, another problem is that the technology to sort and recycle plastic and mixtures of paper and plastic are only beginning to be developed. Several solid waste experts have expressed a concern that the industry construction of the DSD won't facilitate source reduction, reuse techniques, and plastic recycling. Neither the regulation text itself, nor the EHK summary, nor DSD policy mention anything about technology incubators or seed grants to develop plastic recycling technology. However, the GKV and KVG, two plastics recycling councils, may implement another set of licensing fees to fund technological development and market development for plastics recycling.

The EHK summary asserts that the deposit system and packaging licensing can be managed with an easy, comprehensive effort, but this is not obvious. The DSD won't award contracts and is merely an administrative and licensing organ. As Dr. Wollny comments, "The DSD itself will not recycle one kilogram of packaging."

Krey, Wollny, and others have also commented that verification of the recycling rates will be virtually impossible. Either a vast bureaucracy can be created to check on business recycling rates, or businesses must be relied on to accurately report their performance. The regulation requires that 50% of all packaging by weight be recycled between 1993 and 1995. By 1995, 72% of glass, bi-metal, and aluminum, and 64% of cardboard, paper, plastics, and mixed materials must be recovered and recycled. The figures come from multiplying the rate of collection with the rate of separation and reuse; for example, 90% of glass might be collected and 80% recycled, rendering a total recovery of 72%.

As ambitious as these numbers sound, Wollny fears that the heavier materials will be recycled and plastic ignored. He also recommends that this regulation or future guidelines should redesign the waste stream so as to make it more recyclable. Large plastic bottles, for example, should uniformly be high density polyethylene (HDPE).

A Breakthrough

In spite of its potential shortcomings, this appears to be revolutionary legislation for several reasons: it requires comprehensive packaging recycling; it internalizes the cost of that recycling to the cost of the item, rather than privatizing profit and making disposal cost both the government and the public, and the internalization of cost will be readily visible to consumers; other legislation will copy the Verpackungsverordnung of Germany. There is not yet a comparable piece of legislation in another nation. Citizens in Switzerland, for example, have the option to return some kinds of packaging, and there are deposits on many items. However, the Swiss direct democracy and decentralized legislative process have delayed a similar policy in that country. Oregon and Massachusetts have passed packaging laws, but they are far less inclusive; and some large municipalities, such as Minneapolis, Minnesota have passed appropriate packaging ordinances or bans on polystyrene, for example.

Philosophy of Packaging Recycling

According to Wollny, German Environmental Minister Klaus Topfer pressed for this regulation because he believes manufacturers should be responsible for the disposal of packaging. He feels that the cost of disposal should be internalized into the cost of the product. Presumably, the upshot of this is that once the cost of disposal is moved to the front end of its use, free market innovation will shoulder the greater part of the cost of recycling.

Both critics and advocates of the Verpackungsverordnung say that the green dot and the fee on packaging will raise consumers' awareness about their own responsibilities and options to "precycle," to avoid excessive or unnecessary layers of paper and plastic. Clearly, waste reduction has a cultural component and it may simply take consumer response and creativity to lessen waste— something that can't be legislated. There is some concern that the green dot may seem to make a recommendation about the environmental value of product, and an education campaign may be launched to inform the public about the program.

The worst that can be expected of the Verpackungsverordnung is that the source waste stream is barely reduced but that the percentage of recycling and consumer awareness of packaging are somewhat raised. The best that can be expected is that companies significantly reduce packaging at the source and find that they save money at the same time; that sorting and plastic recycling innovations are stimulated; and that this first comprehensive packaging requirement becomes a model for the rest of the world.

****

GERMAN 'GREEN DOT' LAW MAKES BRITISH SEE RED

By *Mike Holderness*, a London-based writer who covers the European recyclables markets.

Britain's wastepaper dealers are [in 1992] complaining that the bottom has fallen out of their packaging-grade market, blaming a German packaging ordinance known as the "Green Dot" program, which went into effect this spring.

Over-collection of wastepaper from packaging in Germany "is depressing prices as they dump it on the rest of Europe," says Geoff Jones of the British Waste Paper Association.

According to the British Paper and Board Federation, 1992 has seen announcements of closure from 10% of the United Kingdom's waste processing capacity. "Some of that would have stayed open had it not been for the German law," says the federation's Brian Bateman.

Ironically, Germany has reported strong prices for wastepaper. In April 1991, prices for packaging-grade scrap had fallen to zero or negative figures, but German mills are now paying real money for the material, notes Oscar Haus of the German Papermakers Federation. According to Haus, the quantities of paper collected for recycling haven't even changed that much, which shows that "the free market system worked very well," he says.

Haus also says that German imports and exports of wastepaper aren't changing that much—a phenomenon he calls surprising. According to Britain's Bateman, however, in the first quarter of 1992, German exports of wastepaper to all countries rose by 30%, as did exports of waste-based packaging products.

Britain's direct trade with Germany is insignificant—it doesn't even show up separately in U.K. figures—but France does a lot of trading with Germany. German exports of paper for recycling to France increased from 197,000 tons in 1989 to 268,000 tons in 1990, and to 454,000 tons in 1991. Meanwhile, wastepaper prices fell; the average declared value across all grades decreased from 328 French francs (FF) per ton in 1990 to FF220 in 1991 (from $68 to $45, according to late August [1992] exchange rates).

This collapse of wastepaper prices in Western Europe dates back at least three years. If one believes the Germans, this is explained by market forces, aided by a powerful environmental movement or moral obligation to recycle. If one believes the British, the Green Dot law represents an indirect subsidy to the German waste processing industry.

The German law specifies, in essence, that all shipping packaging must be recycled. Manufacturers and distributors must either take back their cartons themselves or subscribe to the "Duales System," under which they pay the Duales company a royalty to put a green dot symbol on each package. Waste collectors claim a fee from Duales for each package collected, and the British industry estimates that this operation adds up to about 330 deutsche marks ($190) per ton.

EC Does Little To Help

On the face of it, this might look like a barrier to free trade within Europe; such barriers are outlawed under European Community (EC) law. Members of the European Commission "listen sympathetically, but say the situation is unstoppable," Bateman says.

The EC has its own directive on packaging, the fifth and final draft of which was published in July. Less stringent than the German law, it would require member states to set up "systems guaranteeing the return of used packaging and/or packaging waste." Ten years after the directive's approval, 90% of packaging materials would be required to be "removed from the waste stream for the purpose of recovery."

The difference between this and the German law is that in the EC directive, "recovery" may include incineration to produce energy or waste-derived fuel. According to Haus, the German Papermakers Federation would prefer this to the country's current law, but there is still "psychological" resistance to incineration in Germany, he adds. "We have a plant which opened here in Bonn a few weeks ago, 25 years after the city council first proposed it. Environmental groups went to the highest courts in their attempt to stop it," he says.

It's anyone's guess when the EC directive might be passed, but it will probably be years before certain national governments legislate to implement it. The Germans, Dutch, and Danes have called it too weak, and the British are opposed in principle to anything so interventionist, so they cannot be expected to speed its progress.

France is already considering a new law on waste, offering its recycling industry the "level playing field" with Germany that the British industry wants. Neither of the two British paper associations is keen on legislation. "The market will look after itself," says one industry spokesperson. Both groups, however, have expressed an interest in seeing the British government force an increase in landfill charges.

Outside observers might think that the British are simply whining again—blaming the effects of a deep national recession on external factors. After all, British wastepaper collectors, speaking off the record earlier in 1992, were blaming low prices on the level of exports from the U.S. In fact, however, less than 36,000 tons of U.S. wastepaper reached the U.K. in 1991, and both British paper trade associations acknowledged that the effect on European prices was largely mythical.

There could be even more confusion in store; on January 1, 1993, the second part of the German packaging law takes effect. From then on, not only must trade packaging—such as the cases that hold boxes of cornflakes—be recycled, but consumer packaging—the boxes of cornflakes themselves—must be recycled. This could wreak even more havoc on the shaky European wastepaper markets.

****

CANADA AND FRANCE ESTABLISH THEIR OWN MODELS FOR RECOVERING PACKAGING WASTE

By *Jennifer A. Goff*, managing editor for *Recycling Times*, Washington, D.C.

While the tide of criticism continues to wash over the German Green Dot program, countries in both the European Community (EC) and North America are trying to find their own pieces of legislative dry ground.

For Canada, the most recent development in a global sea of packaging proposals is the Canadian Industry Packaging Stewardship Initiative (CIPSI). "[CIPSI] began with a careful study of other countries' environmental policies," says Derek Stephenson of Resource Integrated Systems (Toronto). There are aspects "to this program that distinguish it from other programs," he adds.

For one thing, CIPSI has a strong provincial, rather than a national, thrust. "The jurisdictional responsibility rests with the individual provincial level. It's highly unlikely anything national will occur unless there is provincial consensus," Stephenson says.

"The agreements in each province—and supporting legislation—should be as similar as possible," explains John Hanson, executive director of the Recycling Council of Ontario (RCO, Toronto, Ontario). "You can't have a million different requirements."

The program, originally developed by the Grocery Products Manufacturers of Canada (GPMC, Don Mills, Ontario), was proposed as a method of reducing packaging waste by 50% by the year 2000, through a system based on economic incentives. In essence, the program is a "demand-pull, market-driven strategy" that places heavy emphasis on market development, according to Stephenson.

Another critical element of CIPSI is the implementation of a well-defined, two-phase transition period. "There are two phases to this program which really distinguish it from other programs," Stephenson says. "There is a *need* for a transition period. It is impossible to go immediately to full internalization of costs."

In Phase I of the program, brand-owners would contribute funds on a per-ton basis for **all** packaging material, regardless of the type of material used. However, if secondary-content packaging material is utilized, the brand-owner can receive a rebate. "So [the program] has an immediate incentive to drive markets for the use of that material." Stephenson says. In addition, there are also rebates for the recovery rates, which have been achieved for each material.

In Phase II of the program, levies will be added to each material to cover the full cost of handling that material. These levies "are quite variable ranging from the levy that will be made on aluminum on a per-ton basis, to what it would be on a PVC [polyvinyl chloride] bottle," Stephenson explains.

An entity entitled the Canadian Industry Packaging Stewardship Organization (CIPSO)—comprised of brand owners, distributors, and importers—would receive funds from the levies and channel the revenues, for the most part, to the municipal funding program.

CIPSO "would directly fund local municipalities to offset the cost of their multi-material collection systems. Municipalities would receive funds according to a 'top-up' formula," according to a statement prepared by GPMC. For example, in each individual province, "a municipal operating standard is established. If the cost of recovering any packaging material is less than the cost of collecting that material minus a voided material cost, CIPSO pays the difference," Stephenson says.

Unlike Phase I in which the calculations were based on the average, in Phase II, the calculations are based on the specific type of material. Consequently, the onus is on the brand-owner to select a material "being fully cognizant of what it will cost to manage that material after its use," Stephenson says.

Markets, Markets, Markets

"The most important element of the program is market development," Stephenson says. "The market development strategy distinguishes it from other programs to date."

The goals of the market development approach are to:

- Increase material revenues to offset the costs of the brand-owners
- Reduce costs throughout the system

- Assess, in Phase II, each material based on the cost of managing it throughout the system
- Drive down the cost of collection and processing
- Invest in nonpackaging uses for these materials
- Use essential rebates
- Move toward true cost accounting and to credit that toward recycling costs
- Increase the revenues received by municipal program operators
- Establish partnership relationships with local, provincial, and federal governments in the investments of those market development funds

"Brand-owners also have very direct ways they can reduce costs," Stephenson says. "Increase secondary content in their packaging; switch to a lower-cost material; and continue to lightweight the material.

"We believe that the advantages of this model over other possible product stewardship approaches is that . . . there are strong incentives to reduce costs through the system."

Wading Through the Details

Though Hanson emphasizes that RCO generally supports CIPSI, the council has issued official recommendations and comments on the initiative: ". . . the RCO acknowledges that the proposed model is not yet fully developed or refined and certain critical details are currently unavailable."

For one thing, "the Newspaper Publishers Group [is] expected to contribute substantially to the pot of money, and they have not been as proactive as the packaging industry," Hanson explains. "Without their enthusiastic participation, it will be difficult to sell this to the municipalities."

And despite the threat of the levy, RCO has expressed concern as to whether or not simply avoiding the levy is sufficient "financial incentive" to encourage industry to practice the "3R"—Reduce, Reuse, Recycle—philosophy.

In addition, Hanson says that the extent to which specific funds will be set aside for market development is currently too nebulous. "The majority of the money would flow through to the municipalities, and it's unclear as to how much money would be available for market development and rebates," he says. "The problem is that until there's a really clear sense of how much money will be available to municipalities, it's difficult to do budgets."

Still, Hanson sees CIPSI as a move in the right direction. "We do view [the initiative] as a positive development at a critical time in the evolution of recycling," he says. An "official" proposal surfaced in 1994, but has not been adopted by any of the provinces.

Meanwhile in France

"Come up with a proposal [to manage packaging waste] or else there will be German-style legislation in Europe." So came the threat from France's Environmental Minister in mid-1991, according to Eric Guillon, directeur général for Eco-Emballage (Perret, France), who recently spoke at the National Recycling Congress in Nashville.

Faced with this admonition, Guillon says, French companies were "persuaded" to devise a system whereby free trade within the EC could coexist with packaging regulations that would protect the environment.

"As the 'first kid on the block,' DSD [Duales System Deutschland] sent a shock wave through Europe because of its potential to be seen as the model packaging policy for the entire European Community," Guillon says. "Though its intent is good, DSD had not anticipated serious financial and logistical problems. From an open trade standpoint, DSD and other packaging laws raise . . . important concerns."

In devising their system, the French strove to avoid some of the pitfalls they perceived in the German plan such as loss of market share; the threat to companies' free choice of packaging; limits on industrial countries; and economies of cost in production (which lead to higher costs for consumers).

Instead, "the [French] packaging decree permitted a positive balance between environmental and business goals," Guillon says. According to the French plan, any manufacturer, importer, or store-brand retailer who uses packaging for household, consumer product marketing "shall be required to contribute to, or provide for, the sound disposal of its packaging waste," Guillon explains.

Still, Guillon adds, under the French decree, industry has three options:

- Place deposits on all consumer packaging
- Implement an approved recovery system for its own packaging, or
- Transfer responsibility for packaging recycling to an approved organization (such as Eco-Emballage)

In November 1992, the French government granted authorization to Eco-Emballage "to channel funds to municipalities for packaging waste recycling," Guillon says. "Eco-Emballage's mandate is to manage all types of household packaging," though other specialized waste companies work in conjunction with Eco-Emballage to handle materials such as glass bottles and pharmaceutical items, Guillon adds.

Though Eco-Emballage has a challenging goal to recover 75% of all packaging waste by 2002, incineration is included in the French decree. "Valorization as defined in the French decree . . . means recovering packaging materials for recycling, for clean incineration with energy recovery, for combusting—the main objective being to ban landfilling by the year 2002," Guillon explains.

Who Pays?

"Companies that choose or use packaging must share responsibility and should provide for funding for recycling and sound waste disposal," Guillon says. Placing fees on domestic packaging would be more cost-effective "than taxes imposed unilaterally by government, or deposits placed on all types of packaging," he adds.

To avoid a tax system, consumer production companies contribute funds on a per-unit-of-packaging basis. For instance, a 1-liter bottle maintains an average fee of one centime, or 1/5 of a U.S. cent. Flexible packaging, however, is calculated on a per-weight basis.

"For '93, our budget is expected—and will be—$80 million," Guillon says. "About 75% of Eco-Emballage's budget will go to municipalities for market price supports."

In addition, Eco-Emballage will contribute to the additional expense of recycling on a per-ton basis. These extra costs are defined as "a difference between multi-material curbside collection plus sorting, and normal collecting costs, and incineration."

Eco-Emballage also deals directly with the municipalities "because they are legally responsible for the collection and disposal of household [waste]," Guillon says. "For the next six years, the garbage collected and sorted by municipalities will be guaranteed to be recycled. Municipalities, in return, must sort to the quality standards of the end markets," he adds.

"More and more, European countries are studying our system. Time will tell if we are right," Guillon says. "But we are confident."

****

THREE-YEAR ASSESSMENT OF THE GERMAN GREEN DOT SYSTEM

By *Jennifer A. Goff, Lisa Rabasca*, and *Kathleen M. White, Waste Age* Publications, Washington, DC.

Green Dot Applauded at Annual Conference

Three years after its initiation, and only six months after the rumored collapse of Germany's Green Dot Program, some industry officials hailed the Green Dot system as the resolution to the current waste policy debate in the U.S.

"We [in the U.S.] have failed in respect to source reduction and we have failed in respect to recycling," said Frank Sudol, manager, Newark, New Jersey, Department of Engineering, at the March 1994 U.S. Conference of Mayors/National Association of Counties-Municipal Waste Management Association's Annual Conference held in Washington, D.C. "[The Green Dot] legislation is influencing the need and outcry for passage of similar progressive packaging legislation in the United States."

Armed with waste generation and source reduction statistics and comparisons, Sudol lambasted the U.S. system, which espouses recycling while, at the same time, doing little to decrease overall waste generation. "While recycling in the U.S. is increasing, so is waste generation," he argued. "On the other hand, while recycling in Germany is dramatically increasing, waste generation is going down. Half a million tons less packaging material was used in 1992 than in the previous year."

Sudol's sentiments were echoed by Counsellor Berthold Goeke, the German representative of the Federal Ministry for the Environment, who further emphasized manufacturer responsibility.

"If we really want to achieve integrated environmental protection, a new philosophy is needed in the field of waste management too—a 'cradle to grave' philosophy which makes the producer responsible for the entire life of his product," Goeke said. "Only if people are aware that they have to take back and recycle the products they themselves have brought onto the market will they reflect . . . on how they can make these products as environmentally friendly as possible."

Goeke did, however, outline certain problems with the Green Dot Program that, he said, "were above all financial in nature." Goeke cited difficulties such as "free riders," which are "firms which although they do imprint the green dot on their packaging . . . pay for far less packaging than they actually produce and than the system has to dispose of." Other financial difficulties Goeke cited included the "collection zeal of the public" which has resulted in "higher costs for collection, sorting, storage, and reuse than was expected," as well as contamination problems when consumers dispose of nonpackaging waste through the dual system.

Goeke briefly touched on the markets dilemma, but relegated his remarks only to plastics. "Substance recycling of plastic packaging was initially very critical," he said. "Due to the collection zeal of the public . . . the quantities collected were greater than the recycling capacities available."

But despite vociferous protests from Germany's European neighbors as well as Asian paper dealers, Sudol denied that Germany was solely responsible for wreaking havoc on foreign wastepaper markets.

"As exports of U.S. scrap paper have weakened, the glut of fiber in Europe—Germany in particular—has been blamed for much of the problem," he said. "While there is a need to increase local and foreign demand, to say that the German program is failing, because it is collecting large quantities of materials is ludicrous."

"To place this figure in perspective, total paper mill utilization of wastepaper [in the U.S.] in 1992 was only 31.2 percent [according to the American Forest & Paper Association's 1992 Paper Mill utilization]," Sudol said. "The USA falls behind 15 other countries in the percentage of wastepaper which is domestically utilized."

Though Goeke extolled the German program as being successful "in particular given the short time involved," he admitted that the program will undergo a "thorough check-up."

"One of the aims of the amended Ordinance is to create a secure basis for investment in so-called feedstock recycling procedures," Goeke said. "A clarification of the term 'substance recycling' is designed to ensure that chemical recycling procedures are also acknowledged as substance recycling."

Also, because of the problem with plastics, Goeke is recommending an extension of the deadline so that the "recycling quota of 64% for plastics does not have to be achieved by mid-1995 but is reduced to a percentage of 60% which has to be achieved by 1 January 1998."

Goeke added, however, that despite the European Community's efforts to prepare a common packaging directive for all European members, Germany along with the Netherlands and Denmark would refuse to participate because the "directive falls far behind our expectations."

German Paper Industry Executive Criticizes Green Dot Program

The German Green Dot Program has left the European wastepaper market in disarray, according to a German paper industry executive.

"As the Packaging Ordinance excluded incineration, as well as landfilling, the only remaining option was exportation," said Dr. Klaus Kramer, managing director of C.D. Haupt Papier and Pappenfabrik & Co. KG, a paper and paperboard mill. "The European waste paper market now became entirely distorted," he added, speaking at the April 1994 Wastepaper V conference in Chicago.

In 1991, Germany approved an aggressive packaging law which requires retailers to take back from its customers any packaging sold in the country. DSD, a private organization, was created by businesses to collect, sort, and recycle packaging.

In his remarks, Kramer outwardly criticized the German government and its law. "The pressure to establish the Dual System, . . . with its sole responsibility being the collection and recovery of the sales packaging directly from households, clearly shows tendencies towards monopolization," Kramer said.

One reason the law has caused problems is that the DSD collected more waste than the German government had anticipated, Kramer said. "The German collection quota increased from 47% in 1991 to 50% in 1992, and even 54% of the consumption in 1993, and thus for the first time surpassed the waste paper input quota, which rose from 50% to 53% of the paper and board production," Kramer said. "The consequence of the paper consumer being today approximately 30% above paper production is a huge oversupply."

As a result of this oversupply, prices for mixed wastepaper dropped to negative numbers, Kramer said.

"A logical consequence of this dramatic price distortion for waste paper was that, parallel to the low demand for packaging papers, prices fell dramatically from mid-1992 on," Kramer explained. "Within two years, prices for waste based corrugated papers fell by 40%, those for various board grades fell by 30%. Accordingly, the price level for corrugated and board packaging collapsed; furthermore, sales of these packaging materials dramatically declined."

Since the 1991 ordinance was passed, packaging used in Germany has decreased by more than 1 million tons, according to Wolfgang Schutt, managing director of INTEC GmbH, a consulting group that helped implement the Green Dot system.

The volume of household garbage also has declined, Schutt said, and Germany expects a sharp decrease in the cost of recycling. However, the cost of packaging has increased significantly in Germany. The average household is paying an additional 160 deutsche marks annually for products contained in packaging collected by the DSD, Schutt admitted.

German citizens may pay even higher prices for products if a proposed amendment to the packaging ordinance is approved. A proposal has been made to introduce a collection debt for those who use transport packaging, such as producers of consumer goods, Kramer said. According to the proposal, these suppliers are to collect their used transport packaging at the receiver's site, mostly at supermarkets, and to bear the incurred costs themselves.

"A second dual system for the removal of transport packaging waste from the consumer's site might be established—the exact opposite to a free market system," Kramer said. "The current disposal channels would be destroyed, uncertainty for all participants in the paper chain as well as rising costs would be the inevitable consequences."

Inform Concludes U.S. Can Learn From Green Dot

A new report released by INFORM, Inc. (New York City), concludes that Germany's solid waste law, specifically its 'polluter pay' principle, is producing some positive results and suggests that certain policy mechanisms—though not all—from the law can be considered as a basis for similar waste reduction strategies in the U.S.

The report, *Germany, Garbage, and the Green Dot: Challenging the Throwaway Society,* was released April 11, 1994, at a press briefing in Washington, D.C., which featured a panel of policy makers and recycling industry professionals offering their views on the potential applicability of Germany's law in the U.S. The briefing was co-sponsored by the Environmental and Energy Study Institute (Washington, D.C.).

The report, which was authored by INFORM's Municipal Solid Waste Program Senior Fellow Bette Fishbein, is being touted by INFORM as the first thorough and authoritative examination of Germany's law, the Ordinance on the Avoidance of Packaging Waste, to be written in English. The Green Dot system, as the law has come to be known, took effect in 1991.

Acknowledging the wide range of responses Germany's ordinance has elicited, Fishbein started out on a cautionary note. "Policies of this type [the German policies] are not thumbs up or thumbs down," she said. "It's a complex policy of which we can take bits from."

Still, she added, Germany's revolutionary 'polluter pay' principle, which has shifted the responsibility for waste management from the public sector to the private sector, is having a far-reaching impact. "The importance of what happens in Germany goes well beyond Germany," Fishbein said. "Almost every country in Europe has similar legislation, or is about to propose similar legislation . . . . Even the recalcitrant Great Britain [has something in the works] now."

Addressing the successes as well as the failures of the ordinance, Fishbein said that since its implementation, Germany has experienced a steady decrease in the amount of packaging used in the country. In addition, companies are now designing more for recycling and are setting up disassembly plants and recycling industries.

The failures of the ordinance, however, have been infamous, she said. Perhaps the most publicized has been the fact that the Duales System Deutschland (DSD), the organization set up by German businesses to monitor Green Dot fees, has been on the brink of bankruptcy. In addition, consumers have been putting non-DSD materials in recycling bins, causing significant contamination, Fishbein said. And, finally, the program has collected far more packaging than it expected to collect and far more than it can handle, she added.

Thus, Fishbein concluded by suggesting that although certain policies from Germany's law can be duplicated in this country, the exact implementation of these policies should not be copied by the U.S. "It's very important to separate the concept of the 'polluter pay' principle from the mechanisms of impact," she offered. "We must address these policy concepts on their own. We can set up our own recycling rates, for example, based on [some of] their concepts."

Offering a municipal perspective, David Gatton, senior environmental adviser for the U.S. Conference of Mayors (Washington, DC), said cities are under the gun through rising mandates and goals to increase recycling. "We project that the average cost to collect and recycle is between $175–200 per ton [of material]. It's expensive and local governments are paying for it," Gatton said. "We're interested in seeing a nationwide debate on [manufacturers responsibility]," he said. "Does anyone feel that you're responsible for the packaging waste in your product when you go to the store?" he asked the audience in conclusion.

Gatton's position was refuted by Tom Rattray, associate director of packaging, worldwide, for Procter & Gamble Co. (P&G, Cincinnati), who said the principles of 'polluter pay' and Germany's ordinance are flawed. "Although we're very active in it, we don't necessarily subscribe to it," he said.

Specifically, Rattray said, Germany's ordinance redefines property rights through its DSD fee system; its definition of pollution is flawed; it gives the public an incentive to cheat because they are under the notion that recycling is free; and it incorrectly suggests that any single sector can manage our solid waste. Referring to Germany's manufacturer responsibility system, Rattray said although his company is for manufacturer responsibility in concept, "we do what we know best, which is to work with recycled material and not [manage] it." For these reasons, a similar ordinance would not work in the U.S., he said.

Instead, Rattray suggested, by making the cost of recycled products comparable to virgin-material products by, among other things, stopping all subsidies on virgin materials, the recovery and reuse of materials will increase. P&G currently uses post-consumer material in its products.

Eliminating virgin-material subsidies is a market development principle that the Recycling Advisory Council (RAC) of the National Recycling Coalition (NRC, Washington, DC) has been working on, said Edgar Miller, director of policy and programs for RAC. "Basically we're dealing with a dysfunctional marketplace right now," he said.

According to Miller, NRC and the RAC "are very supportive of the goals of the German system." Members of both entities believe some sort of national policy on recycling is necessary in this country, he said. "We have attempted to look at the German problem . . . to come up with our own solution," Miller added.

In terms of setting up a national recycling policy similar to Germany's in the U.S., Jim McCarthy, senior analyst for the Congressional Research Service (Washington, DC), said this country is not anywhere close to doing so. Legislation formulated in 1992 that introduced the concept of manufacturer responsibility for packaging was struck down, and legislation on the Resource Conservation and Recovery Act is not up for reauthorization until 1997, he said.

Still, McCarthy said, "my thought is that the U.S. is increasingly alone among industrial countries that do not ask industry to take some responsibility for its packaging."

9 CASE STUDIES

BROOKHAVEN, NEW YORK—NEW ENGLAND CRInc.

By *Christina Thoresen*, former editorial assistant for *Waste Age* Publications, Washington, D.C.

The Town of Brookhaven, New York's $9.5-million materials recovery facility (MRF), owned by the town and operated by New England CRInc. (Chelmsford, MA), has put up some solid recycling numbers in its first three years: The facility processes 400 tpd of commingled recyclables and paper products. Last year, the MRF processed 21,857 tons of recyclables, collected from Brookhaven's more than 415,000 residents.

The MRF's design, however, also has a lot to do with its continued success. In a unique partnership, the Brookhaven MRF's permitting, design, and construction were brought about by the teaming of CRInc. with a civil engineering firm known as The Maguire Group (Foxboro, MA).

Brookhaven went through an elaborate investigation of proposals, and following that, an interview process of the bidders' experience and scope of quality in building and equipment. After that extensive evaluation, the town selected the CRInc./Maguire team.

"[The Maguire Group] was looking for a leader in the materials recycling and systems supply," says Michael Kelly, vice president of resource recovery and recycling operations for the Maguire Group. "After doing our research, we identified CRInc. and [its use of] Bezner technology, as the best qualified company to succeed. We approached CRInc. on a team arrangement and bid the Town of Brookhaven, N.Y., MRF [that way]," he explains. "[For the Brookhaven MRF bid,] our team put in a lot of front-end work," he adds.

It was an unusual contract for the team. "[This] is really a new philosophy in the materials recycling business—where one party is responsible for the design, construction, and operation of the facility," says Kelly. "The Brookhaven [MRF] represents one of the first full-service materials recycling facilities in the country."

Facing Problems Head-On

The MRF has been operational since January 1991—after a 13-month design, construction, and installation schedule that included a few "bumps" along the way. Before construction began, the design team had to meet one problem head-on—the state of New York's newly implemented (at the time) DEC Part 360 Permit (the contract was awarded to the team December 31, 1988).

"We were blazing new trails as to the requirements of development of the permit of a constructed MRF," recalls Kelly. "We did things a little out of sequence, and the town signed on prior to having the permit in place, anticipating a fair amount of time would go into adopting regulations."

A DEC Part 360 permit requires: a contingency plan outlining actions to be taken in case of emergencies; an engineering report describing the facility's construction and operation; an operation and maintenance manual (a day-by-day operational guide describing start-up to closure); and a comprehensive recycling analysis of the facility, identifying the means by which the town would achieve a 50% reduction in waste disposal, as directed by the state of New York.

"The town actually took some risk, and based on [preliminary plans prepared by Maguire] gave us notice to proceed. We were developing permit documents at the same time as [we were developing] engineering documents," Kelly recalls. The DEC Part 360 permit prepared by the Maguire Group for the Brookhaven MRF was the first of its kind to be successfully obtained in the state of New York, according to the company.

Another problem surfaced when Suffolk County, New York, passed stringent regulations to protect groundwater, recalls Kelly. "The county was concerned with any potential liquid waste that could be inadvertently delivered to the [MRF] and seep into the aquifers below."

As a result of Suffolk County's water requirement, the facility was designed with no floor drains in it. "They are of limited use anyway," adds Kelly, "[and] often get clogged, very quickly becoming inoperable."

Instead of floor drains, the facility was designed with a 4-inch sill at all doorways that runs around the perimeter to contain any spillage on the floor. It was also proposed that cleaning the facility be done with portable steam cleaning machines to lessen effluents and water, while a wet vac be used to vacuum up any additional water—enhancements that satisfied Suffolk County's water requirements.

A potential processing problem was also taken care of at the beginning of the design stage. "We included large roll-up doors 20′ wide and 27′ high," explains Kelly. "A number of vehicles delivering materials to the facility are the hydraulic-lift, gravity-dump type, and we had to provide an opening to prevent any accidents that might occur if the lift was not lowered before exiting," he said.

Construction of the facility began in a pre-engineered, industrial-type building. The building was capable of being customized to design plans, and provided a large, clear-spanned area for tipping and receiving materials, as well as for processing and storage.

MRF Processing

Located on a seven-acre site set back from the road, the 40,000-square-foot Brookhaven MRF is "well-buffered from any visual impacts of residents traveling the roadway," Kelly says. The MRF is laid out in such a way, he says, so that materials are constantly moving through the facility.

Processing recyclables begins with trucks dumping all commingled recyclables onto the tipping floor, where the recyclables go through a pre-sort inspection. All of the materials are processed entirely by Bezner technology, with which CRInc. has an exclusive licensing agreement. The mixed recyclables—glass bottles and jars, aluminum and steel cans, and polyethylene terephthalate (PET) and high-density polyethylene (HDPE) containers—are loaded onto a receiving pit pre-sort platform and continue on a conveyor belt past an overhead electromagnetic belt, which extracts ferrous and bi-metal containers, depositing them on a feed conveyor to a ferrous dedicated baler.

The aluminum, glass, and plastics proceed onto a Bezner BSM-60 screening machine, which screens out mixed broken glass and fine materials. The mixed broken glass is captured through a system-wide broken glass recovery system. The flow of materials then divides into two streams, allowing a greater flow of recyclables to be processed at one time.

Bottles, aluminum cans, and plastic containers pass over Bezner's patterned inclined sorting machine which separates aluminum and plastics from the heavier glass that tumbles down. Plastics and aluminum are diverted to a BSM-10 screening machine, where large plastic containers are automatically channeled to the plastics sorting station. Bezner's eddy current system separates aluminum cans from the remaining plastic for processing.

Any additional glass fragments are screened out by a Bezner BSM-10 machine. The glass, in one stream, proceeds to the glass processing line for color separation. Once separated, the glass travels to color-specific glass processors, where it is crushed and stored in concrete bunkers.

Plastics, automatically separated on the plastics sorting belt, are sorted by resin type at Bezner's "head on" sorting stations. PET is separated out for processing in a PET perforator and baler; HDPE goes through a negative sort, and continues directly to an HDPE-dedicated baler for processing.

Paper, which arrives at the facility separately from the commingled recyclables, is manually sorted into four grades of materials: newsprint, corrugated containers, and high- and low-grade paper. Contaminants are removed at the sorting platform and once separated, the paper is stored in bins until baling by a Bollegraaf (Clifton, NJ) baler.

Baled materials are loaded out of a total of eight dock doors, and put onto vehicles for transport to market. Markets are primarily regional, according to Hal McGaughey, CRInc.'s vice president of marketing and business development. "However, we do make use of Canadian and export markets for paper products."

No Need To Worry About the Weather

Workers also benefit from the facility's design. In the processing area, there are roof-mounted supply units to introduce air while suppressing dust, and employees work in enclosed sorting stations. There are three sorting stations: residue—a two-man station for removal of any residual materials that could be harmful to the commingled process; paper—paper grade sort; and commingled—plastic and color-glass sort.

These enclosed sorting stations, which are all environmentally controlled, represent an important step taken to improve the standards for working conditions and sorting efficiency, Kelly says.

The Town of Brookhaven has been very diligent in their planning process, Kelly says. The facility is part of what he calls a solid waste complex—the town operates a landfill, a MRF, and is now in the process of procuring a "dirty MRF"/transfer station to store waste prior to transferring it to a waste-to-energy facility.

For the future, the MRF's design allows for expansion of up to 600 tpd of processed materials. Planned expansions should take place in the third quarter of 1994, says McGaughey, which include expansion of the tipping floor and an enhanced facility sorting capability that will allow both additional materials and additional volume. "Our goal," adds Kelly, "is to keep the MRF fully operational while meeting [those] expansion requirements."

****

DALLAS, TEXAS—COMMUNITY WASTE DISPOSAL, INC.

By *Kathleen M. White*, senior editor for *Waste Age* Publications, Washington, D.C.

When Community Waste Disposal, Inc. (CWD, Dallas), opened for business back in 1984, there were only a handful of waste hauling and recycling companies in the Dallas-Fort Worth metropolitan area. Today, however, the market for haulers and processors has become more competitive, with the number of companies vying for business in the region more than tripling. Still, despite this climate, CWD, which is a relatively small private hauling and materials processing company, has managed to survive and even prosper. Indeed, as the waste management and recycling industry in the area has grown, CWD, too, has grown with it, becoming the largest privately owned waste collection service in the area. And, by remaining a relatively small-scale operation, they are disproving the notion that, in order to be successful, being bigger is better. For CWD, being different is better.

Establishing Collection Contracts First

Although CWD is celebrating its tenth anniversary this year [1994], it wasn't until January 1993 that the company opened their materials recovery facility (MRF) and entered the recyclables processing business. Prior to the opening of the MRF, CWD was primarily a hauling company, transporting municipal solid waste and, in 1989, office waste paper from private and commercial accounts in the Dallas-Fort Worth area. As their business grew, however, and the mood of the country favored recycling, they knew they would have to get involved in collecting and processing recyclable materials, explains Greg Roemer, president, founder, and part owner of CWD.

The company's first foray into this end of the waste management field came with a contract to collect and process materials from apartment dwellers and single-family-home residents in nearby Euless, Texas. Finalized in July 1992, the 72-month contract calls for CWD to collect and process recyclables from approximately 9,000 apartment units and 8,800 single-family houses in the city. According to Roemer, the contract represents the first public/private municipal agreement in the state of Texas in which all residents of a community have access to recycling on their own property.

Although CWD was successful in securing the contract with Euless for a number of reasons, including a competitive price bid, one of the most important was its ability to effectively handle the collection of materials from both apartments and single-family homes, Roemer says. "Our experience with collection, flexibility, and personalized service in efficiently handling recyclable materials from these two sources was the key," Roemer adds.

Since this initial contract, CWD has steadily and quite rapidly built up its clientele base one community—or building—at a time. Within months of the company's contract with Euless, CWD also established contracts to collect and process recyclables from residents of apartment and condominium buildings in Dallas. The contracts, which are established with apartment owners and management companies, are part of what CWD calls its Wash It 'N' Toss It^{R} recycling program. As part of the program, CWD charges the management companies a mandatory rate of $0.49 per unit, per month, for weekly door-to-door pickup of residents' recyclables. The program benefits not only residents who are eager to recycle, but the management companies as well, Roemer says. Management companies receive lower waste collection bills as a result of the program, he points out. As of

April 1994, approximately 1,500 apartment units in the city of Dallas are participating in CWD's program.

In addition to a multitude of apartment residents, CWD is currently collecting and processing materials from single-family homes in a number of communities in the Dallas-Fort Worth area. After the company's contract with Euless and the start of the Wash It 'N' Toss It program, the cities of Double Oak, The Colony, Oak Point, Kaufman, Bartonville, Fairview, and, most recently Frisco, all contracted with CWD for recyclables collection and processing services. "That brings our customer base to approximately 79,000 citizens and growing," Roemer says. Indeed, CWD recently received city council approval for collection and processing contracts for three more Dallas-Fort Worth metropolitan communities.

Besides a buildup of residential recycling customers, CWD has also seen an increase in the number of businesses contracting for recycling services. According to Roemer, CWD currently manages more than 200 office paper recycling programs in the Dallas-Fort Worth area. The company, which, like its residential program, provides collection and processing services for the businesses, is serving some prominent industries in the area, Roemer says. Current businesses CWD has contracted its services with include: GTE World Headquarters; Texas Instruments; Allstate Insurance; United Parcel Service; J.C. Penney; and Mary Kay Cosmetics.

Up and Running in a Hurry

As rapidly as CWD was building up its clientele base, its MRF was coming on line. Altogether, the construction and start-up phase of the 20,000-square-foot facility took approximately six months to complete. The expeditious development of the MRF was due, in part, to the fact that the company was lining up customers in a hurry, Roemer says. While the facility was in its final stages of completion, CWD even had to store materials collected from Euless residents until the MRF was able to process them. In addition, since CWD's MRF is a relatively uncomplicated processing system, the demands on installation time were not as great, he adds.

"Our MRF is very traditional in its capabilities," Roemer continues. Indeed, the facility relies on manual sortation as its primary processing method. Five sorters and a foreman are required to move materials through a typical eight-hour shift. In addition, processing equipment that is currently being used at the MRF is minimal. Besides a Bobcat (Fargo, ND) to move materials on the MRF floor, the facility has an inclined conveyor and a downstroke baler, explains Jim Perry, safety manager for CWD. "At our current tonnage level, this is all the equipment that we need to process materials right now," Perry says.

According to Roemer, the MRF is processing an average of 93 tons per month (tpm) of residential recyclables, although the facility is capable of processing up to 500 tpm. In addition, CWD is processing approximately 331 tpm of mixed office wastepaper at the facility, he says. Processing these different commodities is done intermittently, Perry adds, with roughly three days of the week reserved for processing mixed wastepaper and the remaining two days for the residential recyclables.

Both incoming residential recyclables and office wastepaper arrive at the MRF in commingled form. CWD utilizes a blue bag system of collection for its municipal contracts, so all recyclables arrive in 13-gallon plastic bags. Once the residential recyclables are removed from the bags, they travel up the incline conveyor and past the manual sorting station where they are separated into respective streams.

Currently, CWD accepts traditional household commodities such as aluminum, tin, four colors of glass, old newspapers (ONP), high-density polyethylene (HDPE), and polyethylene terephthalate (PET). In addition, the facility also accepts polyvinyl chloride, low-density polyethylene, polypropylene, and other miscellaneous plastic bottles. "We take all plastic bottles for processing except number six, but we don't talk about numbers to our customers" Roemer says. "We say, if it's a plastic bottle, we'll take it." CWD is the only company in Texas that accepts such a wide range of plastics for processing, Roemer adds.

Once all the materials are separated at the sorting station, they are sent to individual bunkers where they are stored for later shipment. According to Roemer, CWD bunkers as much as 200 cubic yards of each material at a time. When the bunkers are full, he says, the materials are sent to one of two places: directly to an end market or to an intermediate market for baling and further storage. Currently, with the exception of the blue bags, CWD is not baling any materials on premises.

"The incoming material at the MRF is too much for downstroke baler, but not enough to warrant a horizontal baler," Roemer says.

"When we reach our capacity of approximately 1,000 [tpm] then we will purchase a horizontal baler," Perry adds. "For now, however, it's more economical for us to operate this way."

Operations will more than likely change as CWD takes on more customers, Roemer says. Fortunately, although the MRF was built in a hurry, the planning of the facility was certainly not rushed. "We built the MRF with expansion capabilities in mind," Roemer says. "Aluminum separation equipment, for example, can be easily added to our system and we can expand on that, as well."

"We put a lot of equipment in the facility up front anticipating growth," adds Perry. "And we do expect to grow."

Staying a Step Ahead

One of the reasons CWD has been successful with its operation to date is its willingness to go the extra mile servicing customers, Roemer says. The company's decision to collect and process such a variety of plastics, for example, is more a move to stay a step ahead of its competition, than an economic one. Building a better "mousetrap" is the company's philosophy regarding its decision to accept such nontraditional plastics, Roemer says. "We want to beat our competitors and stay ahead in the business," he says. "Accepting materials that your competitors don't is a good way to do it."

So far, CWD has been doing so at the company's expense. "We're probably making less money by collecting these plastics," Roemer says. CWD, which markets all the materials it accepts at the MRF, has found local and regional buyers for all its materials—including most of its HDPE and PET and all its blue bags—except the remaining plastics. "We have to go throughout the U.S. to move these plastics, which can be quite costly," he says. Still, Roemer adds, offering collection and processing services for these materials to its customers is worth it.

And, it seems, the company's strategy is paying off. Even though CWD is not making money by collecting and processing the nontraditional plastics, overall the company's profits have been increasing. According to Roemer, CWD's revenues in 1993 exceeded the previous year's revenues by 30%. Revenues in 1994 are expected to follow the same pattern. "I see our 1994 revenues to be 25-30% higher than our 1993 revenues," Roemer adds. In addition, the company expects to expand both its municipal recycling and refuse collection contracts even more in the next year.

"We're going to continue to market our services to municipalities and we will continue to market our services commercially to free-commerce areas," Roemer continues. In addition, he says, CWD is currently discussing the possibility of building a mixed solid waste and recyclables MRF (or, what is commonly known as a 'dirty MRF'), which would tentatively be operational within the next 24 months. More immediately, however, the company plans on adding more materials to its already varied list of recyclables collected from its curbside programs. Old magazines, miscellaneous metals, more office paper, and possibly more plastics are materials currently being considered for acceptance at the MRF, Roemer says.

"We've been operating as a company since 1984, and we're proud of the way we've done things and our success thus far," Roemer adds. "We're proud of the fact that we've never bought another company, we've never been acquired by another company, we've never changed our name, and that the same founders own and operate it," he says. "We're the largest independent hauling and recycling company in the Dallas-Fort Worth metroplex operating in free-enterprise markets, where you have to go out and scrap for business. The fact that we've had no ownership turmoil shows a relatively solid and stable price system and service that we have."

****

TAMPA, FLORIDA—ReCLAIM, INC.

By *Lisa Rabasca*, editor of *Recycling Times*, Washington, D.C.

Without the support of the New Jersey state legislature, ReClaim Inc. (Tampa, FL), would not be successful, says James Hagen, the company's president and chief executive officer. ReClaim recycles asphalt-based roofing scrap into a cold-mix patching material—known as RePave™—which is used to repair potholes. The company has found that the key to its success is working closely with state legislators to develop state regulations.

"The biggest part [of the business] is dependent upon getting an open discussion with the regulatory people in the state, finding out who those people are, introducing the company to them through a letter, inviting them to look at the company, [and] setting up meetings," Hagen says. "It's about a three month process just to get to see the right person.

"We have been working for four years with New Jersey, and we finally are making progress," Hagen says.

Courting New Jersey

While the company's headquarters are in Tampa, Florida, its two operating facilities are in Camden and Kearney, New Jersey. The Kearney facility opened in 1988, employs 15 people, and processes 125 tpd of roofing material, while the Camden site opened in 1991, has 14 employees and processes 75 tpd of roofing material, Hagen says. The Kearney facility is 7,000-square-feet located on a 2.2-acre site, and the Camden facility is 19,000-square-feet located on a 4-acre site.

Indeed, ReClaim has worked closely with New Jersey. "We didn't immediately embrace them," says Guy Watson, bureau chief of New Jersey's Department of Environmental Protection and Energy. "We had a lot of questions. They provided a number of technical information sheets and reports regarding the material."

Last year, New Jersey issued an executive order requiring all state agencies to purchase recycled products when available. According to the executive order, "If there is a competitive recycled product available that is equivalent or superior in performance to a nonrecycled product and the cost is equivalent or no more than 10% higher, then the state agency must purchase the recycled product." The New Jersey executive order specifically mentions recycled roofing products.

"We tried to mention as many recycled products as possible wherever we could," says Guy Watson, "It just so happens that only one company is producing asphalt-based roofing. It would be nice to have competition but so far we don't have that."

The New Jersey Department of Transportation (DOT) tested RePave for 18 months, Watson says. According to Watson, the head of the state's DOT says cold-patch asphalt normally lasts three minutes to three months, while RePave™ lasted 18 months and longer. "Apparently the stuff adheres to the surrounding paving, not as if it was a patch, but as if it was paving material," Watson says.

The Bayonne, New Jersey, Department of Public Works recently completed an independent analysis of the economic benefits of using RePave™, compared with the traditional methods of repairing potholes. Despite the fact that RePave™ costs more than $100 per ton, using the product results in a savings of $55.31 per pothole, according to the study. "Because of the adhesive properties of this new material, the holes filled with RePave™ did not have to be refilled over the course of the winter and, based upon visual inspection of these holes, will not need an application of hot asphalt in the summer," the study concludes.

Watson says he agrees. "It's not cheap, but the fact that you don't have to replace it, makes it less expensive," he says.

"Any customer that has tested [RePave™], has purchased the product," Hagen says.

After its success in New Jersey, ReClaim has begun working with Florida, Missouri, Tennessee, Washington, and the city of Chicago, Hagen says. He adds that he hopes to establish eight to 10 additional facilities by 1998. The plan is to license the process, technology, equipment, and ability to produce and sell the product, he explains.

From Roof to RePave™

ReClaim uses a proprietary, mechanical process to recycle roofing material into RePave™ and ReActs-HMA, a multi-functional hot-mix asphalt modifier. Through a series of reduction machines, the roofing material is reduced in size to anywhere from one-fourth of an inch minus to talcum-powder sized material, Hagen explains. There is no waste and no by-product, he says, adding that asphalt-based roofing material is 99.9% recyclable.

The company gets its material from three sources: residential roofing, commercial roofing, and from plants that produce roofing scrap. In New Jersey, ReClaim has more than 1,400 roofers as clients and more than 100 haulers who bring their materials to one of the facilities or to one of its 20 drop-off sites. Residents also may leave roofing material at the drop-off sites.

Before ReClaim can process the roofing, however, the material must be sorted. In the case of residential shingles, there is very little sorting required, Hagen says. With commercial shingles, it is important to remove anything that is not asphalt-based, such as tar buckets, tar paper, concrete, and metal.

Generally, there are 11 million tons of asphalt-based roofing generated annually in the U.S., Hagen says. In New Jersey, about 600,000 tons are generated annually and about 60,000 tons are recycled annually by ReClaim, he says. Asphalt-based roofing material generally is 3-5% of the municipal solid waste stream, he adds.

RePave™ was designed to out-perform virgin patching material, Hagen says. "We recognized that the virgin content of what manufacturers use to make roofing material is a high value," Hagen explains. "It has a high concentration of hard asphalt and fiber materials. When combined and reduced to a smaller size, it creates a high value material, that if purchased on a virgin basis, would cost 10 to 20 times more in dollars to create an equivalent product."

Tip Fees Are Key

Another reason for ReClaim's success in New Jersey is the state's high tipping fees. To encourage haulers to bring roofing material to ReClaim, the company charges a tipping fee that is about two-thirds less than the local landfill tipping fee, Hagen says. In New Jersey, landfill tipping fees are more than $100 per ton, he adds.

"We believe there needs to be a discount, even with mandatory recycling legislation in place, to ensure that the ReClaim will get raw feedstock," Hagen says.

"However," Hagen says, "as the value and popularity of RePave™ increases, the company will be less dependent on high tipping fees. Within the next year, ReClaim is hoping to be self-sufficient," he adds.

As ReClaim looks at expanding its operations to other areas, it is focusing on states with local landfill fees of $50 or more per ton, Hagen says. Tampa, FL; Oakland, CA; Chicago, Seattle, St. Louis, Connecticut, and the New York metropolitan area are among the places ReClaim is looking to expand, Hagen says. Additionally, the company is considering some provinces in Canada and Mexico, he adds, noting the North American Free Trade Agreement (NAFTA) is one of the reasons. Also, he says, there is a large infrastructure of asphalt-based roofing material in Mexico City.

Give and Take

ReClaim recently signed a letter of intent with a joint venture partner to build a facility in Chicago. However, Hagen says, a number of factors must fall into place before the recycling plant is built. The city's transportation department must test and fully accept RePave™, he says, and the city must identify asphalt-roofing materials as a recyclable. Additionally, he says, the local landfill tipping fees currently are not high enough to provide an economic incentive to help the material flow through ReClaim.

ReClaim is not shy about telling state and local governments what it expects from them. "ReClaim seeks:

- City/county governments to establish meaningful local ordinances to identify asphalt roofing material as a recyclable which can no longer be carelessly discarded in landfills
- State and local governments to crack down on illegal dumping and illegal 'recycling'
- State DOT testing and approval of ReClaim's end recycled products: RePave™ and ReActs-HMA
- State governments to classify asphalt roofing material as a recyclable which should no longer be dumped into a landfill
- Strong consistent state regulatory environment which certifies legitimate recyclers," according to a promotional document from ReClaim

However, the company is willing to give something back to the community, Hagen says. With the establishment of its next recycling facility outside of New Jersey, Hagen says ReClaim will create a trust fund for the host community. Up to $1 for every ton of asphalt-based roofing material collected will be set aside for the community to use for an environmental project, he explains, adding that the community can choose the project.

Profit for the Future

As Hagen looks ahead, he says he expects the company to turn a profit by the third quarter of this year [1994]. Since the company was founded in 1987, it has recycled more than 290,000 tons of material, he says, but it is yet to turn a profit.

In the last two months, most of ReClaim's sales have come from referrals, Hagen says. Aside from New Jersey, the federal government and McDonald's Corporation (Oak Brook, IL) are becoming ReClaim's largest customers, he adds.

Last year, ReClaim manufactured and sold 900 tons of RePave™. In fall 1993, the company upgraded its equipment line. As a result, ReClaim expects to manufacture 3,000 tons of RePave™ in 1994.

"By the fourth quarter of 1995, our product sales should exceed our tipping fees," Hagen says, adding that he hopes ReClaim will be the first recycling facility in the U.S. to turn a profit from its product sales.

ACRONYMS AND ABBREVIATIONS

AFPA	American Forest and Paper Association
ANSI	American National Standards Institute
API	American Paper Institute
BFI	Browning-Ferris Industries
BLS	Bureau of Labor Statistics
Btu	British thermal unit
C&D	construction and demolition
CIPSI	Canadian Industry Packaging Stewardship Initiative
CIPSO	Canadian Industry Packaging Stewardship Organization
C/N	carbon to nitrogen
CPO	computer printout
CTD	cumulative trauma disorder
CWD	Community Waste Disposal, Inc.
DSD	Duales System Deutschland
EC	European Community
EIA	Environmental Industry Associations
EPA	Environmental Protection Agency
ESCR	environmental stress crack resistance
FF	French francs
GDRRA	Greater Detroit Resource Recovery Authority
GPMC	Grocery Products Manufacturers of Canada
HDPE	high-density polyethylene
HHW	household hazardous waste
IPC	intermediate processing center
IRS	Internal Revenue Service
ISRI	Institute of Scrap Recycling Industries
IWSA	Integrated Waste Services Association
MRF	materials recovery facility

MRRF	materials resource recovery facility
MSDS	material safety data sheets
MSW	municipal solid waste
NAFTA	North American Free Trade Agreement
NFPA	National Fire Protection Association
NIMBY	Not in My Back Yard
NOPRP	National Office Paper Recycling Project
NRC	National Recycling Coalition
NSWMA	National Solid Wastes Management Association
OCC	old corrugated containers
ONP	old newspapers
OSHA	Occupational Safety and Health Act
OWP	office wastepaper
PET	polyethylene terephthalate (aka PETE)
PPE	personal protective equipment
PPI	Producer Price Index
PVC	polyvinyl chloride
RAC	Recycling Advisory Council
RCO	Recycling Council of Ontario
RDF	refuse-derived fuel
RSS	Resource Recovery Systems, Inc.
RUMAC	Rubber Modified Asphalt Concrete
SPI	Society of the Plastics Industry
TDF	tire-derived fuel
tpd	tons per day
tpm	tons per month
UBC	used beverage can
WRC	Waste Recyclers Council
WSM	Waste Stream Management
WTE	waste-to-energy

INDEX